Table of Contents

I0503976

Table of Figures

Table of Tables

ACRONYM LIST:

BMI	Buried Mine Identification
BOSS	Buried Object Scanning Sonar
CCD	Charged Coupled Device
COI	Contacts of Interest
CONOPS	Concept of Operations
CPA	Closest Point of Approach
DoD	Department of Defense
EO	Electro-Optic
ESTCP	Environmental Security Technology Certification Program
FAU	Florida Atlantic University
FOV	Field of View
GPS	Global Positioning System
MCM	Mine Countermeasures
NOMWC	Naval Oceanography Mine Warfare Center
NSWC PCD	Naval Surface Warfare Center, Panama City Division
NUSSRC	National Unmanned Systems Shared Resource Center
ONR	Office of Naval Research
PMA	Post Mission Analysis
R&D	Research and Development
REMUS	Remote Environmental Measuring UnitS
RI	Reacquisition/Identification
S&T	Science and Technology
SCM	Search, Classify, Map
SERDP	Strategic Environmental Research and Development Program
UUV	Unmanned Underwater Vehicle
UXO	Unexploded Ordnance
WFM	Wideband Frequency Modulated
WHOI	Wood Hole Oceanographic Institution

1.0 INTRODUCTION

1.1 BACKGROUND

There are underwater munitions sites around the world where the quantity and the type of munitions are either unknown or not very well documented. These sites are unsafe and can't be utilized for any other purpose. The sites need to be surveyed and all munitions identified. The National Unmanned Systems Shared Resource Center (NUSSRC) located at the Naval Surface Warfare Center Panama City Division (NSWC PCD) has multiple Office of Naval Research (ONR) Mine Countermeasure (MCM) Unmanned Underwater Vehicle (UUV) systems integrated with advanced sonar, magnetic and electro-optic sensors. These systems provide a capability for experimentation and demonstration of current Science and Technology (S&T) and Research and Development (R&D) programs assets as they apply to the detection and classification of underwater mines to be used for detection, assessment and characterization of underwater munitions.

1.2 OBJECTIVE OF THE DEMONSTRATION

The objective is to leverage the ONR MCM UUV assets to evaluate and demonstrate the applicability of existing ONR MCM UUV technologies integrated with advanced sensor packages for facilitating detection, assessment and characterization of underwater munitions. A demonstration of a survey site was performed to establish current capabilities of the UUV technologies for munitions detection and characterization. The concept of wide area assessment followed by small-area detailed surveys and target reacquisition was demonstrated. The performance of the UUV technologies for the underwater munitions problem was established via Post Mission Analysis (PMA) of all sensor data.

2.0 TECHNOLOGY

2.1 TECHNOLOGY DESCRIPTION

Two specific UUV assets in the NUSSRC inventory that can address the detection and characterization of underwater munitions in an organic fashion are the Remote Environmental Measuring UnitS 100 (REMUS 100) and the Bluefin12 Buried Mine Identification (BMI) UUV systems. These systems and their onboard sensor technologies are described below. In addition, since the data collection event for this experiment occurred during the 2011 ONR MCM S&T Demonstration, an opportunity existed for a third system, referred to as the REMUS 600 BMI UUV, with a separate magnetic sensor, the Laser Scalar Gradiometer (LSG), to collect data over the field too. ONR has agreed to provide this data to us for comparison to the data from the Real-time Tracking Gradiometer.

2.1.1 – Remote Environmental Measuring Units 100 (REMUS 100) UUVs

The REMUS 100 UUV, depicted in Figure 1 with a sample display of its primary data products comprised of sonar and camera imagery, is designed to operate in water depths up to 100 meters with reliable navigation and localization accuracy. The primary data product, medium and high resolution sonar imagery, comes from the integrated Marine Sonic dual frequency sonar system that can operate at both 900 and 1800 kHz and map 10 to 100 meters wide swaths in a single pass depending on the resolution required for the operational area. The sonar data is stored onboard and downloaded upon completion of the assigned missions. In addition to sonar, camera and environmental data such as conductivity, temperature, bathymetry, and optical backscatter information can be collected along with vehicle navigation and operational performance. The REMUS 100 system is approximately 80 inches in length, 7.5 inches in diameter and has an approximate dry weight of 80 to 100 pounds and can be handled by a single operator. The REMUS 100 typically has 4 to 6 hours mission duration at a speed of 3.5 knots. This system can provide rapid wide and small-detailed assessments and characterization of the areas contaminated of proud and partially buried munitions.

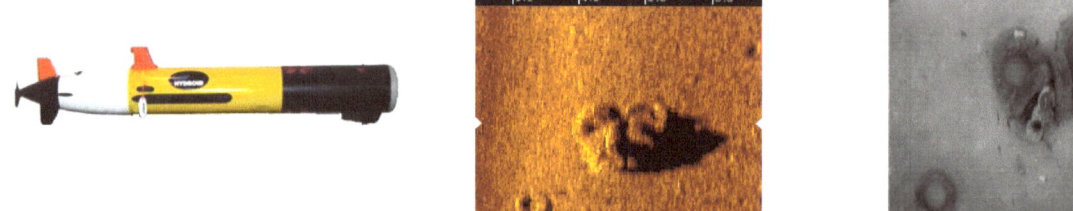

Figure 1. REMUS 100 UUV and sonar/camera imagery products from its 900 kHz side scan sonar and onboard camera, respectively

2.1.2 – Bluefin12 Buried Mine Identification (BMI) System

The Bluefin12 BMI system, depicted in a cut-out view in Figure 2, is based on the Bluefin12 UUV. The Bluefin12 is a 12.75" dia. UUV developed by Bluefin Robotics. This system is integrated with a multi-sensor package that consists of the Buried Object Scanning Sonar (BOSS) system, a passive multi-axis Real-time Tracking Gradiometer (RTG) and an Electro-Optic (EO) sensor.

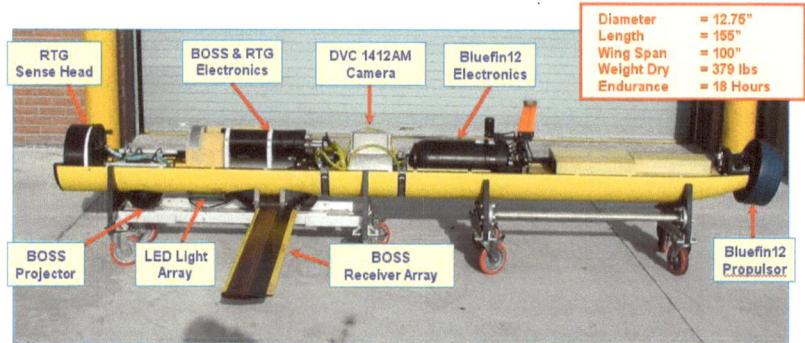

Figure 2. Bluefin12 BMI UUV

The BOSS system is a downward-looking sonar that uses a broad-band; low frequency Omni-directional acoustic projector and 20-element hydrophone receive arrays embedded in each of two 1-meter length wings. Beam forming and synthetic aperture processing generate multi-aspect

Three-Dimensional (3D) imagery of the seabed and provides detection and classification of not only proud but also partially and fully buried Contacts of Interest (COI). The BOSS system is shown in Figure 3. A close-up of the BOSS receiver wings and Omni-directional transmitter as integrated on the UUV is shown in Figure 3a. Figure 3b shows the Bluefin12 UUV on the Launch and Recovery Device (LRD) during field survey operations.

Figure 3. BOSS system as integrated on Bluefin 12 UUV: a) close-up of the BOSS receiver wings and transmitter and b) the Bluefin 12 UUV on the LRD system.

The BOSS system collects and stores raw acoustic backscatter data for each of its 40 elements and navigation/UUV data received from the UUV computer during the survey missions. The data is downloaded from the BOSS computer for PMA upon recovery of the UUV.

For each ping, a 3D matrix of pixels centered at focal points (Xf) on the seafloor is generated via time-delay beam forming; focusing and synthetic aperture processing techniques. This data process generates the multi-aspect 3D imagery. Figure 4 illustrates the BOSS processing scheme for generating the 3D (x, y, and z) matrix of pixels for each ping. The (Xf) focal points are summed at each position in space for all elements correctly taking into account the travel times from the position of the projector and the receiver elements. UUV attitude (roll, pitch and yaw) are taken into account in the travel time calculations. Figure 5 shows the typical anatomy of the BOSS multi-aspect 3D image for both a proud and a flush-buried target. The BOSS multi-aspect 3D image consists of the X-Y (top), X-Z (front) and Y-Z (side) perspective views and where the X-Z and Y-Z views provide target burial depth information. The red lines in the X-Z and Y-Z perspective views are representative of the water-sediment interface and are provided via the Doppler Velocity Log (DVL) sensor or by searching the peak signal from analysis of the backscatter signal. Figures 5a and 5b show proud and fully buried (below the sediment-water interface) targets, respectively.

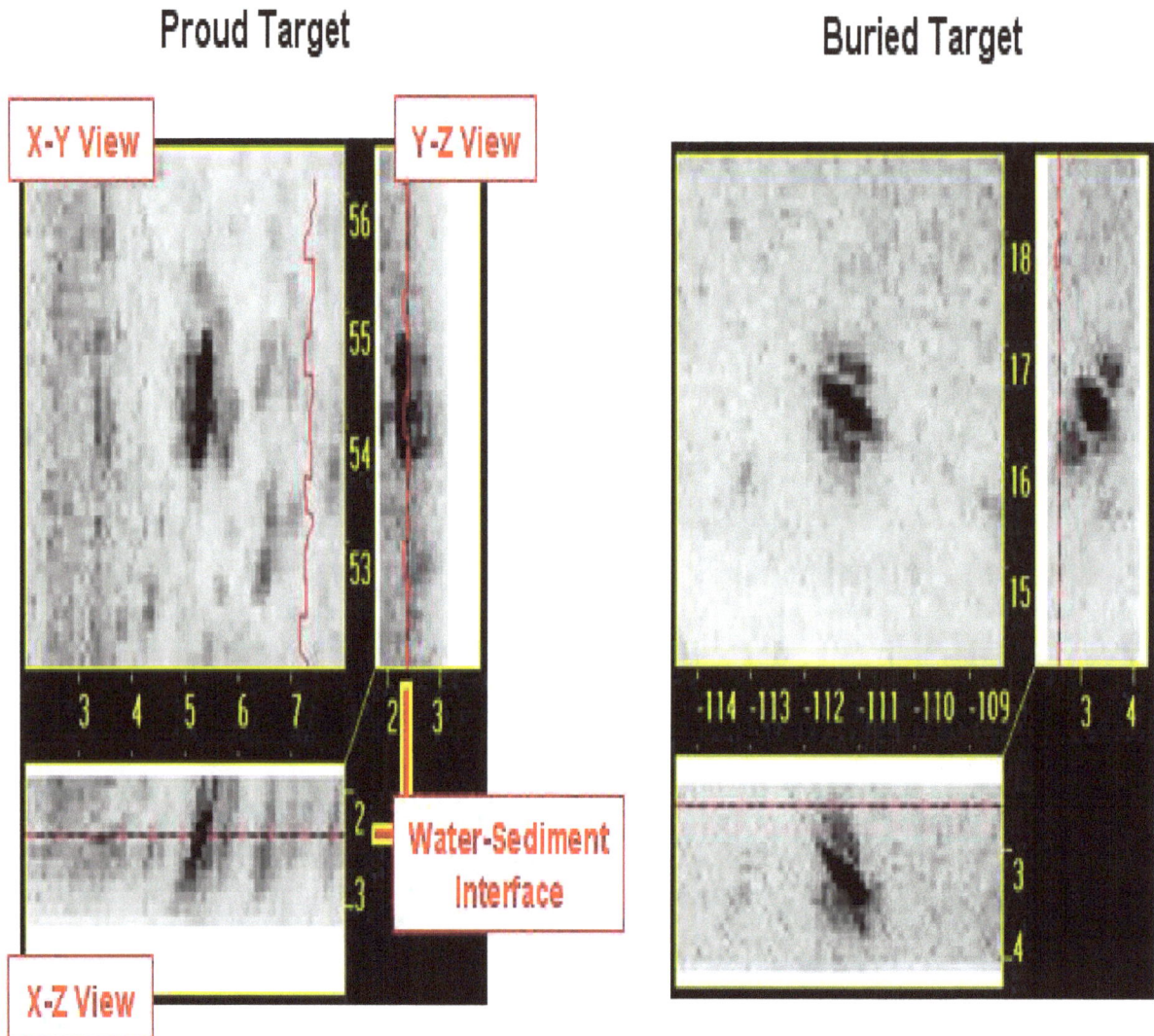

Figure 5. Anatomy of BOSS 3D multi-aspect imagery: a) proud target and b) fully buried target.

The RTG is a small passive magnetic sensor using fluxgate magnetometers measuring 3-orthogonal magnetic-field vector components at 3 spatially separated points in space. The RTG sensor is encased in a watertight housing and is integrated onto the nose section of the Bluefin12 UUV. The data collected from the magnetometer channels is used to synthesize six tensor gradient components of the magnetic field of which five components are independently processed to extract the magnetic moments and three-dimensional position of a ferrous target. A software application calculates the target localization in meters relative to the Closest Point of Approach (CPA) and co-registers with BOSS and EO sensor target localization with a mean absolute cross range difference of 0.41 meters. The target magnetic moment is calculated in amp-m^2 from gradient data. The RTG sensor was a significant aid in discriminating ferrous/nonferrous targets and clutter during Autonomous Unmanned Vehicles (AUV) Fest 08 archeology surveys. Figure 6 shows the RTG magnetic sensor in both an open (showing the fluxgate magnetometers) and enclosed state (mode for integration onto UUV) and its data product showing the calculated time series magnetic gradients of a target.

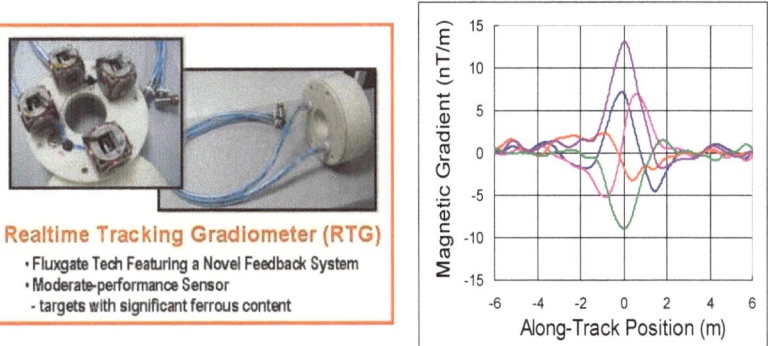

Figure 6. RTG sensor and data product displaying time series target gradient data

The EO sensor is based on the DVC 1412AM high-resolution Charged-Coupled Device (CCD) progressive scan interline camera. This sensor provides high resolution camera imagery of the seafloor and of any targets within its Field-of-View (FOV) when water conditions are permissible. The recorded format of the data is raw pixel data that can be post processed to enhance the imagery.

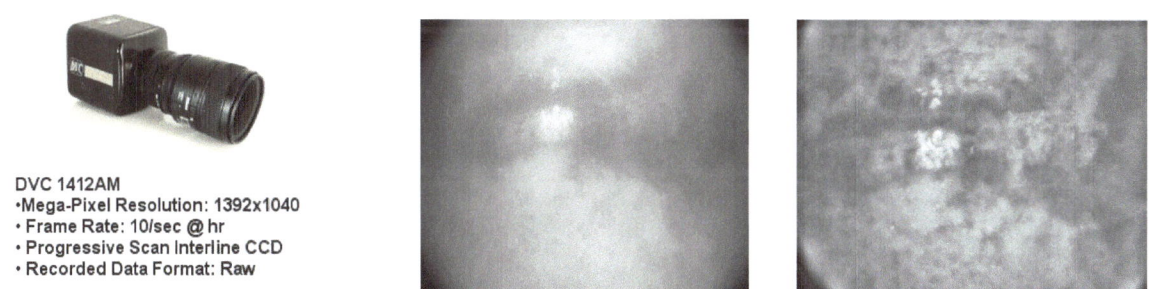

Figure 7. Bluefin12 BMI UUV EO and samples of its data products

The Bluefin12 BMI system usually operates at low altitudes (~ 3 meters from sea floor) to provide for all sensors to detect and localize targets within each of the sensor's FOV. At these altitudes the UUV does not provide a large swath survey capability like that of the REMUS 100 system; however, the Bluefin12 BMI systems does provide sonar imaging and RTG discrimination between ferrous and non-ferrous targets of not only proud but also partially and fully buried targets. The EO sensor provides high quality target images when water clarity is permissible; hence, the EO sensor provides identification capability.

The fusion of the data from the three sensors provides an enhanced capability in the classification and eventual identification of detected targets. Figure 8 illustrates the final data fusion product that is provided by the Bluefin12 BMI UUV. The BOSS and EO images show the co-registered RTG localization. An effort to enhance BOSS target discrimination and identification via post mission acoustic color analysis is presently being investigated.

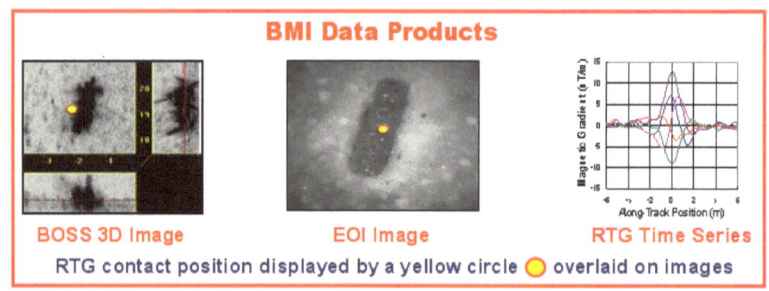

Figure 8. Bluefin12 BMI UUV data product

In 2008 under the Strategic Environmental Research and Development Program (SERDP) MM-1507 project[1], a towed version of the BOSS system that consists of the same Omni-directional transmitter and a receiver array of 160-elements laid out in four rows of 20-elements on each side in similar fashion to the Bluefin12 BMI UUV system integrated with the Bluefin12 BMI UUV RTG sensor conducted survey operations on a target field deployed in the St. Andrew Bay area in NSWC PCD. The target field included rows of real, inert Unexploded Ordnance (UXO) (81 mm and 4.2" mortar projectiles), simulated UXO (5x30cm and 10x30cm steel cylinders), concrete markers (6" diameter x 12 inch long) and assorted debris (concrete clump, tire, aluminum icosahedron, etc.). The rows of inert and simulated UXO were flush buried or buried up to 30 cm deep. The targets 70 in each row were separated by 1 meter. The target field included 4.2" mortar at several penetrating angles (45/75 degrees from horizontal) and piles of several 4.2" mortar flush buried or buried up to 30 cm deep. The concrete markers were deployed vertically and were hemispherically-shaped on the top end to provide acoustic reflection at any lateral aspect angle.

The results obtained from the towed system showed the capability of the 160-channel BOSS system of detecting, imaging and localizing small, proud and buried targets. Analysis of the 160-channel BOSS towed system data processing all channels and then disabling 120 channels to simulate the Bluefin12 BMI UUV system, showed that the Bluefin12 BMI UUV has the capability of detecting, imaging and localizing small targets. Figure 9 shows BOSS multi-aspect 3D images from the 160-channel towed system processed using all 160-channels and using only 40-channels to simulate the Bluefin12 BMI UUV BOSS sensor. Figure 9a corresponds to BOSS imagery of the small, 5x30 cm simulated targets generated by processing all channels in the 160-channel towed system. Figure 9b also corresponds to BOSS imagery of the small, 5x30 cm simulated targets generated by processing 40 channels in the 160-channel towed system to simulate the Bluefin12 BMI UUV BOSS system.

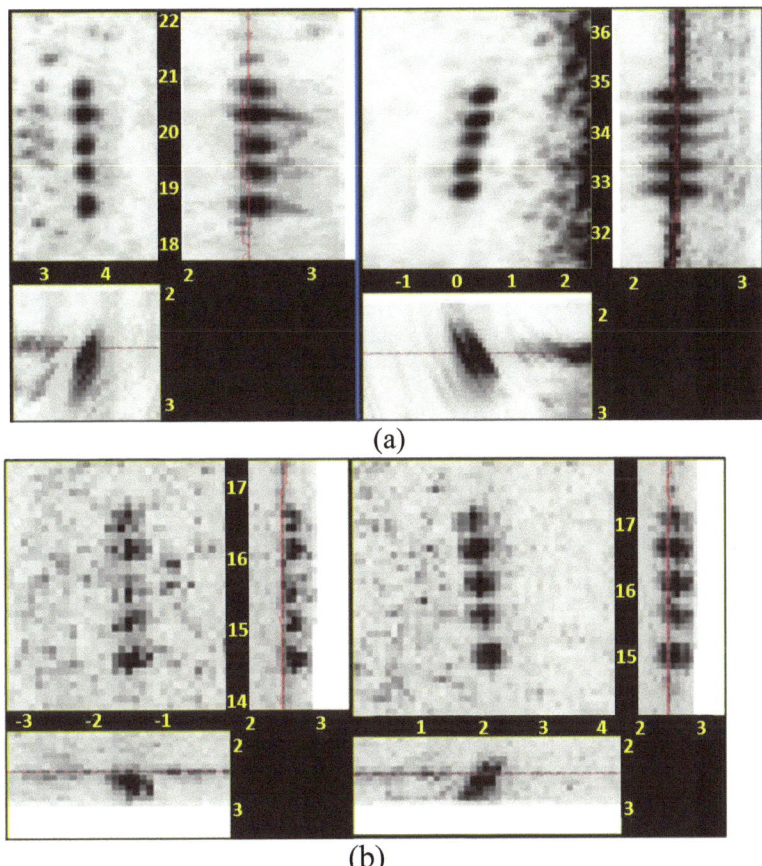

Figure 9. BOSS multi-aspect images generated from the 160-channel towed system: (a) processing all 106-channels and (b) processing only 40-channles to simulate Bluefin12 BMI UUV BOSS sensor. The vertical axis is the along track distance in meters, and the horizontal axis is the cross vehicle track in meters.

2.1.3 – Laser Scalar Gradiometer (LSG)

The LSG is an ultra-sensitive magnetic sensor developed by Polatomic, having high sensitivity relative to the RTG. The Detection, Classification and Localization (DCL) signal processing algorithm is similar to RTG algorithm. The LSG, which utilizes scalar, instead of vector sensing elements, was successfully demonstrated to localize mine-like targets through simulation, land-based tracking and eventually through in-water experiments onboard a REMUS 600 UUV. Figure 10 shows a breakdown of the LSG system.

In addition, with the LSG sensor on board the REMUS 600 UUV there is a Marine Sonic dual frequency sonar system that can operate at both 900 and 1800 kHz to image object features proud to the bottom and an EO sensor which provides high quality target images when water clarity is permissible.

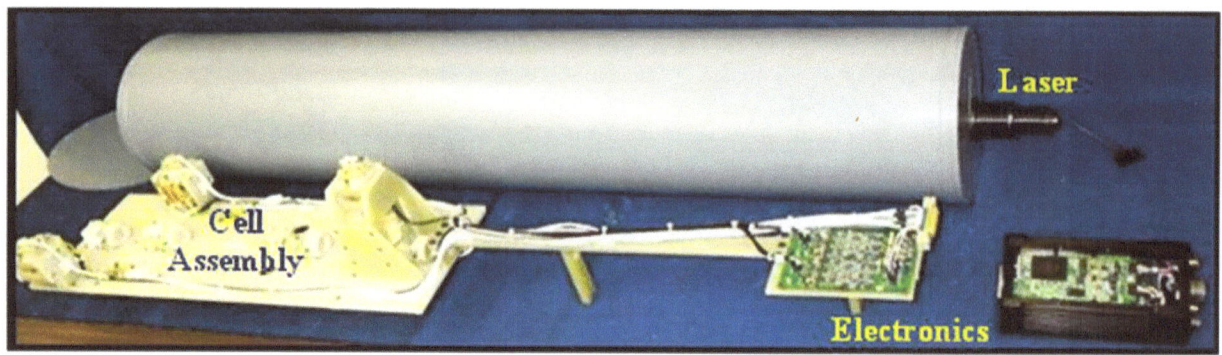

Figure 10. Breakdown of components for the Laser Scalar Gradiometer (LSG) magnetic sensor system

2.2 ADVANTAGES AND LIMITATIONS OF THE TECHNOLOGY

The Naval Oceanography Mine Warfare Center (NOMWC) has used the REMUS 100 UUVs extensively for MCM and harbor surveying exercises. These UUVs have also been successfully demonstrated in ONR-sponsored exercises such as AUV Fest 2007 and 2008 and also at Frontier Sentinel 2010 where the primary function was to survey areas for suspected proud threat mines. The Marine Sonic 900 and 1800 kHz side scan sonar systems integrated on the REMUS 100 UUVs has proven reliable and effective in providing high quality sonar imagery for MCM and survey operations.

The Bluefin12 BMI UUV and its onboard multi-sensor package has been tested extensively from 2004 through 2010 as a Reacquisition/Identification (RI) system in the ONR's MCM concept of employing a Search, Classify, Map (SCM) UUV followed by a RI UUV. This system has also been successfully demonstrated at AUV Fest 2007 and 2008 and also at Frontier Sentinel 2010 where its primary mission was in the reacquisition and classification of mine targets and included fully buried targets. The system has also been demonstrated in detecting and imaging not only proud but also fully buried sea cables.

The primary focus of the systems to be demonstrated has been the MCM mission. Tests and demonstrations of these systems against underwater munitions, which tend to be smaller in size than conventional bottom sea mines are needed to evaluate the applicability of these technologies for detection and subsequent classification of underwater munitions.

The UUV systems have operational limitations. Water environments with high currents (>2 knots) will significantly affect vehicle dynamics and in turn affect sensor performance. Risk of collision is significant if operating in areas where protruding obstructions in the sea bottom are present. High sea states (Sea State >4) will make UUV launch and recovery operations difficult and unsafe. The navigation performance of the UUVs is vital to the survey Concept of Operations (CONOPS) to be executed during the demonstrations. Navigation and Global Positioning System (GPS) errors can cause follow-on surveys to be off the programmed survey mission plan.

Shallow water operations will have some impact in the performance of sonar due to an increase in surface reverberation. The RTG and LSG may not perform as well against smaller ferrous munitions that are somewhat isolated; but may be effective in detecting and localizing clusters of small ferrous munitions, although they will not be able to distinguish individual targets closely spaced together. The water clarity will impact the performance of the EO sensor and cameras.

The raw data from the BOSS test runs can be made available for other researchers, including those interested in exploring the use of acoustic color for munitions detection and discrimination, as long as established classification guidelines from the Office of Naval Research are followed.

3.0 PERFORMANCE OBJECTIVES

3.1 PERFORMANCE OBJECTIVES - Remote Environmental Measuring UnitS 100 (REMUS 100)

The performance objective for the REMUS 100 is to first provide a wide area search over the entire site areas with the 900 kHz side scan sonar. Once the wide area search is complete the sonar imagery from that search will provide information on the overall assessment of proud objects in the area and if any are contacts of interest. A follow-on lower altitude, small-area detailed survey and target reacquisition of the contact of interest will be performed operating the sonar at 1800 kHz. The 1800 kHz frequency will provided with picture quality sonar imagery that enhances classification and identification of the proud contacts of interest. Along with sonar data camera imagery will provide picture quality imageries of the contact of interest.

Unfortunately the REMUS camera module was not available during the test. During a previous unrelated operation the camera module computer developed a problem which caused it to go into a non-recording state. This problem was not corrected in time for the camera to be used during these tests.

Unlike the REMUS camera module problem, the camera was operating on the Bluefin12 BMI vehicle. Unfortunately during the time of the survey low water clarity produced imagery that didn't provide anything of interest.

The LSG survey was performed on a different day and the EO sensor on board the REMUS 600 UUV was able to provide target imagery. At the time of the LSG survey, the water clarity was favorable for target imagery to be collected. Results of the EO sensor imagery are presented in section 7.9 of this report.

3.2 PERFORMANCE OBJECTIVES – Buried Object Scanning Sonar (BOSS)

The performance objectives for the BOSS system are summarized in Table 1. The objectives include assessment of the probability of detection, the degree of burial, size, shape, and orientation determined and localization accuracy against a combination of small and large UXO targets.

The planned tests are not blind tests, but are designed to determine the sidescan, BOSS and RTG combined capabilities against proud and buried targets and to determine whether further optimization of the multi sensor approach, and further performance testing are warranted.

We seek to demonstrate and measure the ability to classify all contacts using: 1) the BOSS to measure target size, shape, and orientation, 2) the RTG sensor to determine which BOSS contacts are ferrous and non-ferrous and to measure the magnetic moment of the ferrous contacts and 3) compare BOSS with REMUS sidescan data to indicate which contacts are buried and which are proud and to enhance localization of the contacts.

Table 1. Performance Objectives for BOSS Sensor

Performance Objective	Metric	Data Required	Success Criteria
Quantitative Performance Objectives			
Detection of all munitions of interest	Percent detected of seeded items	• Location of seeded Items	Pd=0.90
Determine the degree of burial, size, shape, and orientation of contacts.	Percent correct classification of the burial depth, size, shape, and orientation of the seeded targets.	• Validation data for selected targets	Demonstration of >90% correct classification of the size, shape, and orientation of the targets. Length and burial measurements considered correct shall be within 25% of actual.
Location accuracy	Average localization error in meters from latitude and longitude ground truth localization for seed items	• Location of seed items surveyed to accuracy of 1.0m	ΔLocalization Error < 1m from ground truth
Qualitative Performance Objectives			
Ease of use		• Feedback from technician on usability of technology and time required	

3.2.1 – Metric

Compare the number of targets detected by the BOSS system to the number of known targets seeded by divers.

3.2.2 – Data Requirements

BOSS data will be collected over the seeded areas and analysis will follow to identify the number of targets detected by the system. The number of targets detected by the BOSS system will be cataloged and verified against the target listing and ground truth localization provided by the divers.

3.2.3 – Success Criteria

The objective will be considered to be met if more than 90% of the ground truth targets are detected and accounted for.

3.3 OBJECTIVE: DETERMINATION OF TARGET SIZE, SHAPE, AND ORIENTATION - Buried Object Scanning Sonar (BOSS)

Determine the effectiveness of the BOSS sensor in classifying target type by target size, shape, and orientation from the BOSS imagery. Additionally, correlation of data between BOSS imagery with REMUS sidescan images and RTG data will indicate the degree to which contacts can be classified as buried or non-buried, as ferrous or non-ferrous, and by their magnetic moment.

3.3.1 – Metric

Determine the percent correct determination of size, shape, and orientation of all targets contacted by the BOSS that can be correlated to the targets in the planted field.

3.3.2 – Data Requirements

BOSS data will be collected over the seeded areas and analysis will follow to identify and discriminate between targets of interest and non-targets. The number of targets detected by the BOSS system will be cataloged and verified against the target listing and ground truth localization provided by the divers.

3.3.3 – Success Criteria

The objective will be considered to be met if more than 90% of the ground truth targets are detected and accounted for.

3.4 OBJECTIVE: LOCATION ACCURACY – Buried Object Scanning Sonar (BOSS)

Determine the effectiveness of the BOSS sensor for localization of proud to fully buried targets.

3.4.1 – Metric

Compare the BOSS target localization with the diver provided target ground truth.

3.4.2 – Data Requirements

BOSS data will be collected over the seeded areas and analysis will follow to identify and localize the targets of interest. The BOSS localization for each target will be cataloged and verified against the target listing and ground truth localization provided by the divers.

3.4.3 – Success Criteria

The objective will be considered to be met if more than 90% of the BOSS target localizations are within 1.0 meters from the diver reported ground truth localizations.

3.5 PERFORMANCE OBJECTIVES - Real-time Tracking Gradiometer (RTG) / Laser Scalar Gradiometer (LSG)

The performance objectives for the RTG system are summarized in Table 2. The objectives include a demonstration of the performance of detection, the ability to measure the magnetic moment and be able to localize the contacts or clusters of targets. Unfortunately the RTG experienced hardware failure during the data collection event leading to lost data. A likely candidate for this failure is the oil-filled cables that connect the sensor head to the electronics bottle. These cables and their connection points have failed in the past with similar effects on the data. One of the data channels experienced intermittent failure allowing it to be used to construct a gradient for some of the runs. For times when this channel was faulty, a similar gradient could be constructed that would allow the data to be analyzed, albeit at a loss of accuracy and confidence. Unfortunately, a different data channel was completely broken for the entirety of the event and the nature of the gradient constructed using this channel made it impossible to construct a similar replacement gradient. Thus, of the 5 gradients to be used for target localization, only 4 were available and of those 4, only 3 were ideal at all times. Additionally, the collected data was fairly noisy and the signal-to-noise ratio was often very small for these sorts of small dipole targets, making data analysis difficult and the final results poor. A refurbishment and recalibration of the RTG system would be beneficial for future testing events and will be necessary to improve results.

Since the data collection event for this experiment occurred during the 2011 ONR MCM S&T Demonstration, an opportunity existed for another magnetic sensor to operate over the UXO field. The Laser Scalar Gradiometer (LSG) was able to collect data over the UXO field and its results are reported along with the RTG results that occurred in sections 7.7 and 7.8 of this report. The LSG is a more modern sensor than the RTG, based on optically pumped helium cells rather than flux gates and is intrinsically more sensitive. The initial plan for this additional data collection was simply to compare results with the RTG. However, due to the issues encountered with the RTG, it is possible to use the data from the LSG in lieu of the RTG entirely.

Table 2. Performance Objectives for RTG Sensor

Performance Objective	Metric	Data Required	Success Criteria
Quantitative Performance Objectives			
Detection of all ferrous munitions of interest	Percent detected of seeded ferrous items	• Location of seeded Items	Pd=0.90
Measurement of ferrous target magnetic moment	For individual planted targets, the percent of targets whose magnetic moment is measured during a flyover to within 20% of that measured in the laboratory, for the particular altitude flown.	• Validation data for selected targets	Demonstration of >90% of individual ferrous targets moments measured to within 20% of actual.
Location accuracy	Average localization error in meters from latitude and longitude ground truth localization for seed items	• Location of seed items surveyed to accuracy of 1m	ΔLocalization Error < 1m from ground truth
Qualitative Performance Objectives			
Ease of use		• Feedback from technician on usability of technology and time required	

3.5.1 – Metric

Compare the number of ferrous targets detected by the RTG/LSG system to the number of known ferrous targets seeded by divers.

3.5.2 – Data Requirements

RTG/LSG data will be collected over the seeded areas and analysis will follow to identify the number of targets detected by the system. The number of targets detected by the RTG/LSG system will be cataloged and verified against the target listing and ground truth localization provided by the divers.

3.5.3 – Success Criteria

The objective will be considered to be met if more than 90% of the ground truth magnetic targets are detected and accounted for.

3.6 Objective: Measurement of Target Magnetic Moment - Real-time Tracking Gradiometer (RTG) / Laser Scalar Gradiometer (LSG)

Determine the ability of the RTG sensor to measure target moment during flyovers at varied altitudes. The magnetic moment is estimated by statistically fitting windowed time series to a magnetic-dipole model. The properties of the moment, including dipole strength and orientation, are extracted in this process.

3.6.1 – Metric

The percentage of ferrous targets on the signal target lines that are measured to within 20% of actual during flyovers.

3.6.2 – Data Requirements

RTG/LSG data will be collected over the seeded areas and analysis will follow to identify and discriminate between targets of interest and non-targets. The number of targets detected by the RTG/LSG system will be cataloged and verified against the target listing and ground truth localization provided by the divers.

3.6.3 – Success Criteria

The objective will be considered to be met if more than 90% of the ground truth targets are detected and accounted for.

3.7 OBJECTIVE: LOCATION ACCURACY- Real-time Tracking Gradiometer (RTG) / Laser Scalar Gradiometer (LSG)

The effectiveness of the RTG/LSG localization of proud to fully buried ferrous targets.

3.7.1 – Metric

Compare the RTG/LSG target localization with the diver provided target ground truth, and determine the percentage of BOSS detections of seeded targets that can be identified as non-ferrous using RTG/LSG data.

3.7.2 – Data Requirements

RTG/LSG data will be collected over the seeded areas and analysis will follow to identify and localize the targets of interest. The RTG/LSG localization for each target will be cataloged and verified against the target listing and ground truth localization provided by the divers.

3.7.3 – Success Criteria

The objective will be considered to be met if more than 90% of the RTG/LSG target localizations are within 1.0 meters from the diver reported ground truth localizations.

4.0 SITE DESCRIPTION

4.1 SITE SELECTION

The site, named "ESTCP Site 4", is located in the St. Andrew Bay area between Davis and Redfish Point and the main channel in Panama City, Florida. This area consists of a mud and sandy-silt bottom with typical water depths of approximately 12 meters. The ESTCP Site 4 area is depicted in navigational chart snippets shown in Figure 11. The top snippet shows the St. Andrew Bay area from the Hathaway Bridge in the north to just south of the bay entrance from the Gulf of Mexico. The bottom snippet is a zoom of the area encompassing the ESTCP Site 4 area.

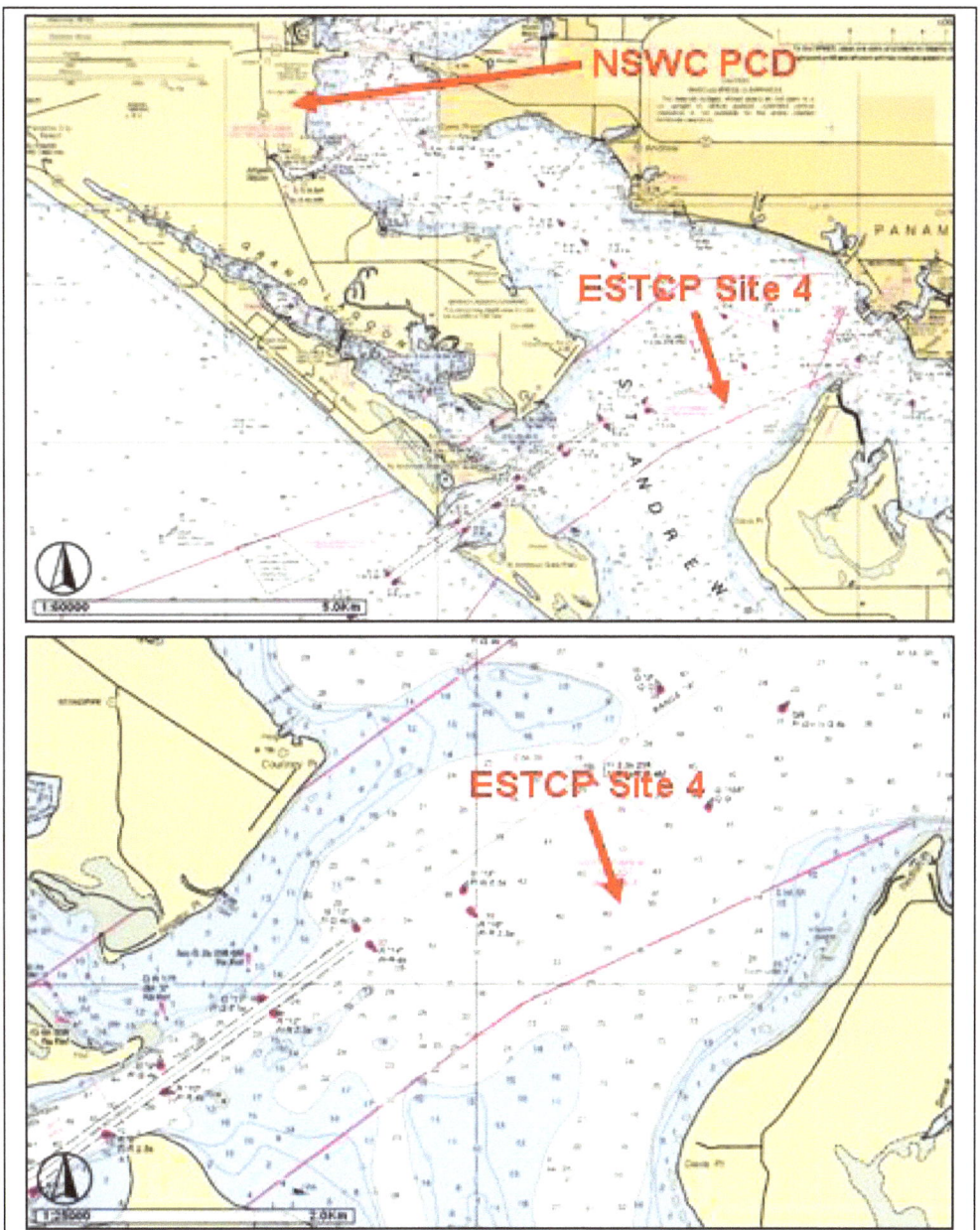

Figure 11. St. Andrew Bay area and location of the ESTCP Site 4 Target field area.

4.1.1 ESTCP Site 4 Target Field Layout

The ESTCP Site 4 area was seeded with proud and buried inert Unexploded Ordnance (UXO) ranging from 50-caliber shells to the larger bomb-sized munitions. Divers provided target positions via plumb line and Global Positioning System (GPS). The target field was marked using five concrete cinder blocks placed on the east and west sides of the field. The layout of the target field is illustrated in Figure 12. The light blue squares, red crosses and dark blue triangles correspond to the cinder blocks, 50-caliber shells, and UXO targets, respectively. The burial states of the targets are designated as F-flush buried or a P-proud. Table 3 provides a listing of the target plumbed positions and burial states (proud, buried). The rows highlighted in green represent the UXO targets. The UXO targets are bullet-shaped with approximately constant shell thickness throughout the length of the body.

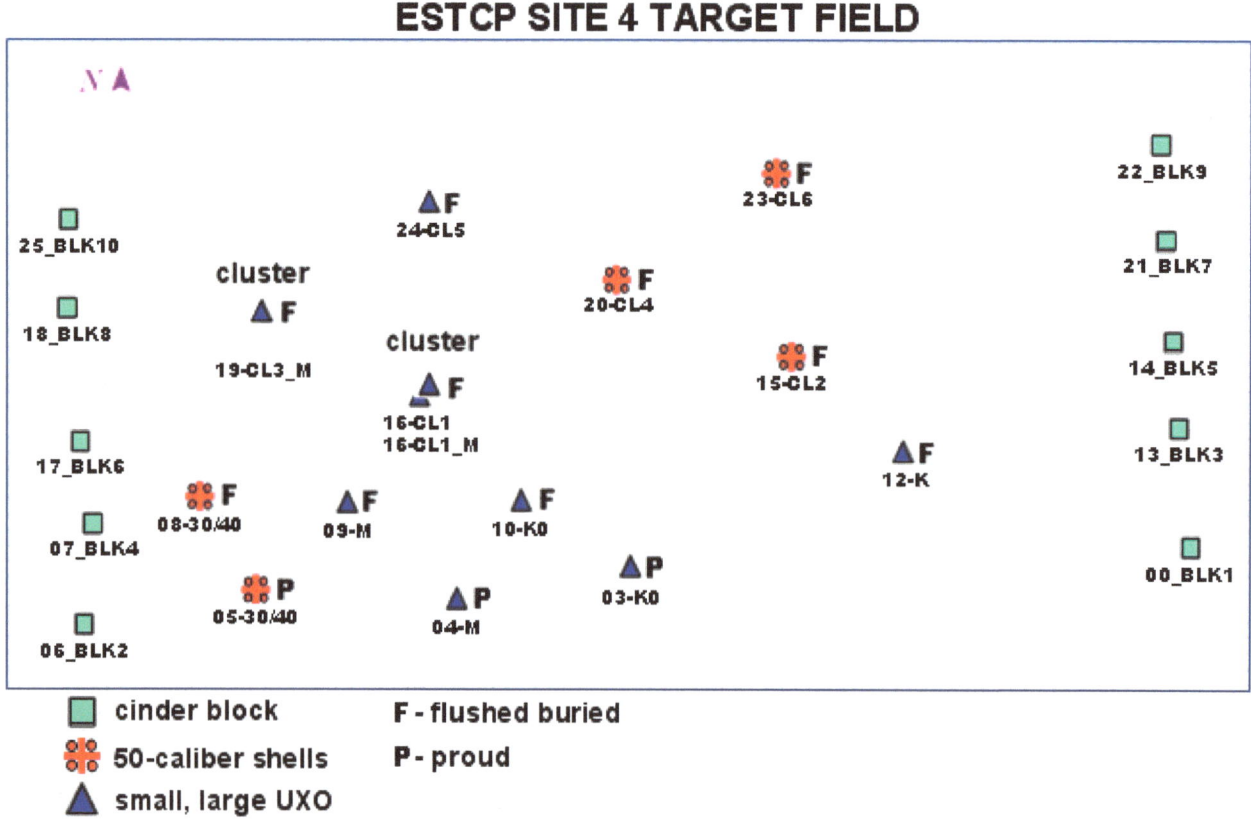

Figure 12. ESTCP Site 4 target description and field layout.

Table 3. Target listing, status and plumbed positions

ID	LAT	LONG	TYPE	WATER DEPTH	STATUS
00_Block1	30 8.24711	-85 41.5652		39	Block - Proud
03-K0	30 8.24615	-85 41.5943	K	38	K - Proud
04-M	30 8.24482	-85 41.6035	M	38	M - Proud
05-30/40	30 8.24509	-85 41.6140		39	10/50 cal - Proud
06_Block2	30 8.24386	-85 41.6231		39	Block - Proud
07_Block4	30 8.24809	-85 41.6226		39	Block - Proud
08-30/40	30 8.24917	-85 41.6169		38	10/50 cal - Flush
09-M	30 8.24886	-85 41.6092	M	38	M - Flush
10-K0	30 8.24898	-85 41.6001	K	38	K - Flush
12-K	30 8.25112	-85 41.5802	K	39	K - Flush
13_Block3	30 8.25218	-85 41.5659		39	Block - Proud
14_Block5	30 8.25589	-85 41.5662		38	Block - Proud
15-CL2	30 8.25500	-85 41.5859		39	10/50 cal - Flush
16 CL1	30 8.25394	-85 41.6051	K	39	K - Flush
16-CL1_M	30 8.25355	-85 41.6053	M	39	M - Flush
17_Block6	30 8.25144	-85 41.6233		39	Block - Proud
18_Block8	30 8.25723	-85 41.6239		39	Block - Proud
19-CL3-M	30 8.25668	-85 41.6139	M	39	M - 4 inch depth
20-CL4	30 8.25834	-85 41.5950		39	10/50 cal - 4 inch depth
21_Block7	30 8.26013	-85 41.5667		39	Block - Proud
22_Block9	30 8.26412	-85 41.5669		39	Block - Proud
23-CL6	30 8.26287	-85 41.5867		39	10/50 cal - 8 inch depth
24-CL5	30 8.26167	-85 41.6050	M	39	M - 8 inch depth
25_Block10	30 8.26099	-85 41.6239		39	Block - Proud

4.2 SITE HISTORY

The ESTCP Site 4 is in an area where survey testing has been conducted in the past. Once testing is completed, the targets are recovered. If any target is not recovered at the completion of a test, a record of the last know location of the target is recorded. As for small munitions use in these sites, there has not been any used in the past. If a target remains, it will be of a larger type munitions which will be very easy to distinguish. The larger target will be considered as clutter that could very well be present in any field.

4.3 SITE GEOLOGY

This site is in about 10-14m of water and is close to the shoreline of the bay. This area consists of a mud and sandy-silt bottom.

4.4 MUNITIONS CONTAMINATION

Small munitions have not been used in either of the proposed test areas in the past. Any munitions-like targets that are present will likely be large and are expected to be distinguishable using the BOSS and RTG/LSG sensors.

4.5 DESIRABLE SITE CHARACTERISTICS FOR A POSSIBLE SECOND DEMONSTRATION

The current UUV system as configured can operate effectively in sandy and muddy environments with water depths of 10 meters. The system has been operated in 5 meter water depths, however, the sonar field-of-view is reduced due to an increase in surface reflections. Because the UUV operates at 3 knots, water currents must be taken into account. Water environments with high currents (>2 knots) will significantly affect vehicle dynamics and in turn affect sensor performance. Risk of collision is significant if operating in areas where protruding obstructions in the sea bottom are present. High sea states (Sea State >4) will make UUV launch and recovery operations difficult and unsafe. Desirable site characteristics for a second demonstration like those surveyed in the Panama City operating areas of the Gulf of Mexico and St. Andrew Bay would be recommended to confirm the contributions of each sensor.

5.0 TEST DESIGN

5.1 CONCEPTUAL EXPERIMENTAL DESIGN

A small underwater munitions test site was planted in the Davis Point area of St. Andrew Bay near NSWC PCD. Five nominal 100m long target lines spaced 2m apart was laid parallel to each other. The first two lines had individual targets spaced along the lines and aligned at random orientations relative to the target lines. One line had the targets proud and the other line the targets were flush buried. The third through fifth lines had two clusters of targets on each line. The beginning and end of each target line was marked with a proud cinderblock. Lines one and two provide baseline measurement data for each individual target type. Lines three through five provided acoustic and magnetic data for clusters of munitions in different states of burial and orientation. The planted locations of the targets were determined by their drop locations from the surface vessel using plumb lines and a GPS system with Hypack software having an accuracy of less than 1m error when conditions are right. Typically, a line is tied to the target and once the target is deployed, the line is pulled tight up to the surface and a GPS reading at that position is obtained. A problem with this method is that the surface craft will drift and the actual GPS reading may not be directly over the target; this error will also increase in deeper waters.

The data collection involved four phases: The first phase was a wide area search over the entire site areas using the REMUS 100 with 900 KHz side scan sonar. This provided a map of proud sonar contacts. PMA was performed on the 900 kHz sonar imagery to both assess the surveyed areas and identify contacts of interest. The second phase was a follow-on lower altitude, small-area detailed survey and target reacquisition performed with the REMUS 100 operating the sonar at 1800 kHz. At the 1800 kHz frequency operators are provided with picture quality sonar imagery that enhances classification and identification of proud contacts. REMUS 100 camera imagery was not collected because the camera module was unavailable during the test. In the third phase, the Bluefin12 BMI sensor suite scans the field at a high altitude (5m) for a wide area scan, to allow the BOSS sensor to image both proud and buried contacts. At this altitude the BMI EO and RTG sensors did not provide useful data, however the combination of BOSS and REMUS side scan data discriminated between buried and proud contacts. The fourth phase performed a small area surveys at low altitude with the Bluefin12 and full BMI sensor suite over each contact detected in the earlier phases. The direction of scan over all elongated contacts was made parallel to the contact's axis to provide the best BOSS imagery.

5.2 SITE PREPARATION

The site was planted with targets to represent a munitions field, with targets ranging from large series bombs to small munitions including bundles of 10-50mm rounds and casings. Taking advantage of other test field laying events for the 2011 ONR MCM S&T Demonstration to reduce project costs, some targets had been placed prior to the final completion of the field layout. Placement of the final targets was completed the 13th and 14th of June of target clusters and targets planned for burial. Figure 12 shows the current field layout. Single targets were placed at random lateral orientations to the track axis. Each cluster was placed with targets in random lateral and vertical orientations, distributed within a 3m diameter circle.

5.3 SYSTEM SPECIFICATION

5.3.1 – Buried Object Scanning Sonar (BOSS)

The Buried Object Scanning Sonar (BOSS) is an unique bottom-looking sonar that generates 3-dimensional, multi-aspect imagery of fully buried, partially buried and proud targets. It features a Wideband Frequency Modulated (WFM), Omni-directional projector that transmits pulses currently in the band from 3 up to 20 kHz and hydrophone receive arrays embedded in two 1-meter length wings that measure the backscattering from the seabed. The BOSS system provides typical temporal, across-track and along-track resolutions of 8, 40 and 7 centimeters at a range of 10 meters. The hydrophone outputs are digitized, matched-filtered and coherently summed using time-delay focusing to form an image of the seabed. The focused data are stored in a 3D matrix where matrix indices correspond to focal points under the seabed. Maximum intensity projections of the 3D matrix data onto orthogonal planes form three views of the seabed. The maxima of a sequence of overlapping projections form a multi-aspect image projection. The sonar operator uses the three orthogonal multi-aspect image projections to view top and side views of buried targets.

5.3.2 – Real-time Tracking Gradiometer (RTG)

The Real-time Tracking Gradiometer (RTG) is comprised of three flux gate vector magnetometers arranged in a triangular pattern and used to measure the magnetic gradient created by any magnetic targets (e.g. most metallic targets). The RTG is located in the nose of the vehicle, primarily to isolate it from intrinsic magnetic noise.

5.3.3 Alternate Magnetic Sensor – Laser Scalar Gradiometer (LSG)

The LSG is a more advanced sensor than the RTG, based on optically pumped helium cells rather than flux gates and utilizes scalar, instead of vector sensing elements, makes the LSG intrinsically more sensitive. The LSG is integrated onboard a REMUS 600 AUV developed by Wood Hole Oceanographic Institution (WHOI).

5.4 CALIBRATION ACTIVITIES

5.4.1 – Buried Object Scanning Sonar (BOSS)

Following the REMUS 100 runs, the next test runs will use the BMI suite to collect data against the two lines containing spaced individual targets. This will provide baseline data for each target type in a horizontally laid, proud and flush buried configuration. These runs will be made parallel to the target lines. Correlation between BOSS data and documented information will confirm proper system operation.

5.4.2 – Real-time Tracking Gradiometer (RTG)

The same runs as in 5.4.1, over the two individual target lines, will provide baseline data for the RTG. As the locations of the different targets will be documented and the target dipole strengths can be computed, the locations and strengths as determined by the RTG can be checked in order to determine the accuracy of the sensor. Localization of a dipole target within <1m is considered acceptable. Accurate determination of dipole strength is not considered as a criterion for acceptance.

5.4.3 Alternate Magnetic Sensor – Laser Scalar Gradiometer (LSG)

The same procedure stated in 5.4.2 for the RTG will be performed with the LSG.

5.5 DATA COLLECTION PROCEDURES

5.5.1 – Buried Object Scanning Sonar (BOSS)

The UUV operator plans and programs the UUV surveys over the area of interest. The BOSS system features a wideband frequency-modulated (FM), omni-directional projector that transmits pulses currently in the band from 3 up to 20 kHz and 40-channel hydrophone receive arrays embedded in two 1-meter length wings that measure the backscattering from the seabed. Raw acoustic backscatter data from the hydrophones is collected and stored in an onboard hard drive during the survey missions. BOSS raw data is downloaded upon recovery of the UUV.

5.5.2 – Real-time Tracking Gradiometer (RTG)

The RTG records data at 125 Hz with a simple list of the magnetic field values as seen by the 12 channels. This data is stored on the onboard computer and is downloaded along with the navigation data for post-processing.

5.5.2 – Laser Scalar Gradiometer (LSG)

Magnetic data is processed onboard the sensor and once the survey is completed localization data snippets are downloaded via WIFI. This is an autonomous, embedded capability for the LSG to localize targets in-stride and to extract acoustic and optical images from the others sensors on board and co-registered that data with LSG localizations, permitting analyst review in-stride during the completion of the surveys.

6.0 DATA ANALYSIS PLAN

6.1 PREPROCESSING

6.1.1 – Buried Object Scanning Sonar (BOSS)

During PMA, the hydrophone outputs are digitized, matched-filtered and coherently summed using time-delay focusing to form 3D image projections of the seabed and targets. The BOSS playback application generates maximum intensity projections of the 3D matrix data onto orthogonal planes to form three views of the seabed. The maxima of a sequence of overlapping projections form a multi-aspect image projection. The sonar operator uses the three orthogonal multi-aspect image projections to view and analyze top and side views of buried targets.

6.1.2 – Real-time Tracking Gradiometer (RTG)

The raw RTG data is first processed from the raw form into a column of magnetic values for the 12 different channels, and then down sampled to be combined with the navigation data. The next step is to apply motion compensation to the magnetic data since sensor motion through the Earth's magnetic field will generate noise. Finally, dipoles are fit to overlapping sections of data, approximately 20 seconds in duration or roughly 40m in length.

6.1.3 – Laser Scalar Gradiometer (LSG)

The data processing is very similar to the RTG method whether operating in real-time on the vehicle or done in post processing. The process is based on the exact same analysis used on the RTG data.

6.2 TARGET SELECTION FOR DETECTION

6.2.1 – Buried Object Scanning Sonar (BOSS)

During PMA, the sonar operator uses the three orthogonal multi-aspect image projections to view and analyze top and side views of targets. Measurements of length and diameter will be used to select targets of interest since these parameters are known.

6.2.2 – Real-time Tracking Gradiometer (RTG)

The final processing step attempts to fit a magnetic dipole to small segments of data. A confidence value is generated indicating the quality of the fit. High confidence values indicate a target has been detected while low confidence values indicate no target is present in that segment of data. Known targets are to be used in testing, allowing the detected locations and dipole strengths to be verified.

6.2.3 – Laser Scalar Gradiometer (LSG)

The target selection for detection is the exact same process used for the RTG sensor.

6.3 PARAMETER ESTIMATION

6.3.1 – Buried Object Scanning Sonar (BOSS)

The operator will analyze the BOSS images and measure the length and width of targets, target localization (latitude and longitude) and burial depth form the cases where a buried target is encountered. The ability to measure the target dimensions, localization and depth is a function within the BOSS playback application.

6.3.2 – Real-time Tracking Gradiometer (RTG)

The observed signal is linear with respect to dipole moment and non-linear with respect to dipole distance of separation. An iterative non-linear optimization algorithm is used to find such a dipole position that minimizes in least squares sense the difference between observed and calculated signals. The algorithm allows for single or multiple dipoles to be modeled. The effect of estimated dipoles on reference channels is estimated and removed via a feedback loop.

6.3.3 – Laser Scalar Gradiometer (LSG)

The signal processing is based on a variant of the codes used to analyze the RTG data.

6.4 DATA PRODUCT SPECIFICATION

6.4.1 – Buried Object Scanning Sonar (BOSS)

The final output from the BOSS playback application will be in the form of BOSS imagery of the targets of interest with the measured target parameters (length, width, burial depth and localization). Analysis of the BOSS imagery and the use of the measured parameters will provide the data to determine target detection and discrimination and target localization accuracy.

6.4.2 – Real-time Tracking Gradiometer (RTG)

The final output from the RTG will be a simple list of target locations, dipole strengths and moment vector that can then be compared against the sonar / EO data.

6.4.3 – Laser Scalar Gradiometer (LSG)

The data product is exactly the same as from the RTG.

7.0 PERFORMANCE ASSESSMENT
7.1 Remote Environmental Measuring UnitS 100 (REMUS 100)

Figure 13 shows the REMUS 100 UUV mission over the test field. The mission was a combine 900 kHz and 1800 kHz sonar surveys. The blue lines with arrows, represents the track lines and vehicle direction as it traveled though the test field surveying at 900 kHz. The purple lines with arrows, represents the track lines and vehicle direction as it traveled though the test field surveying at 1800 kHz. The green (+) symbol represents the locations of the targets detected. Sonar imagery of this is displayed in Figures 14 – 26. The red (+) symbol represents the location of a detected UXO target out of place from Cluster 16-CL1_M. It is thought that the UXO target was moved by a shrimper's nets as they went through the test field.

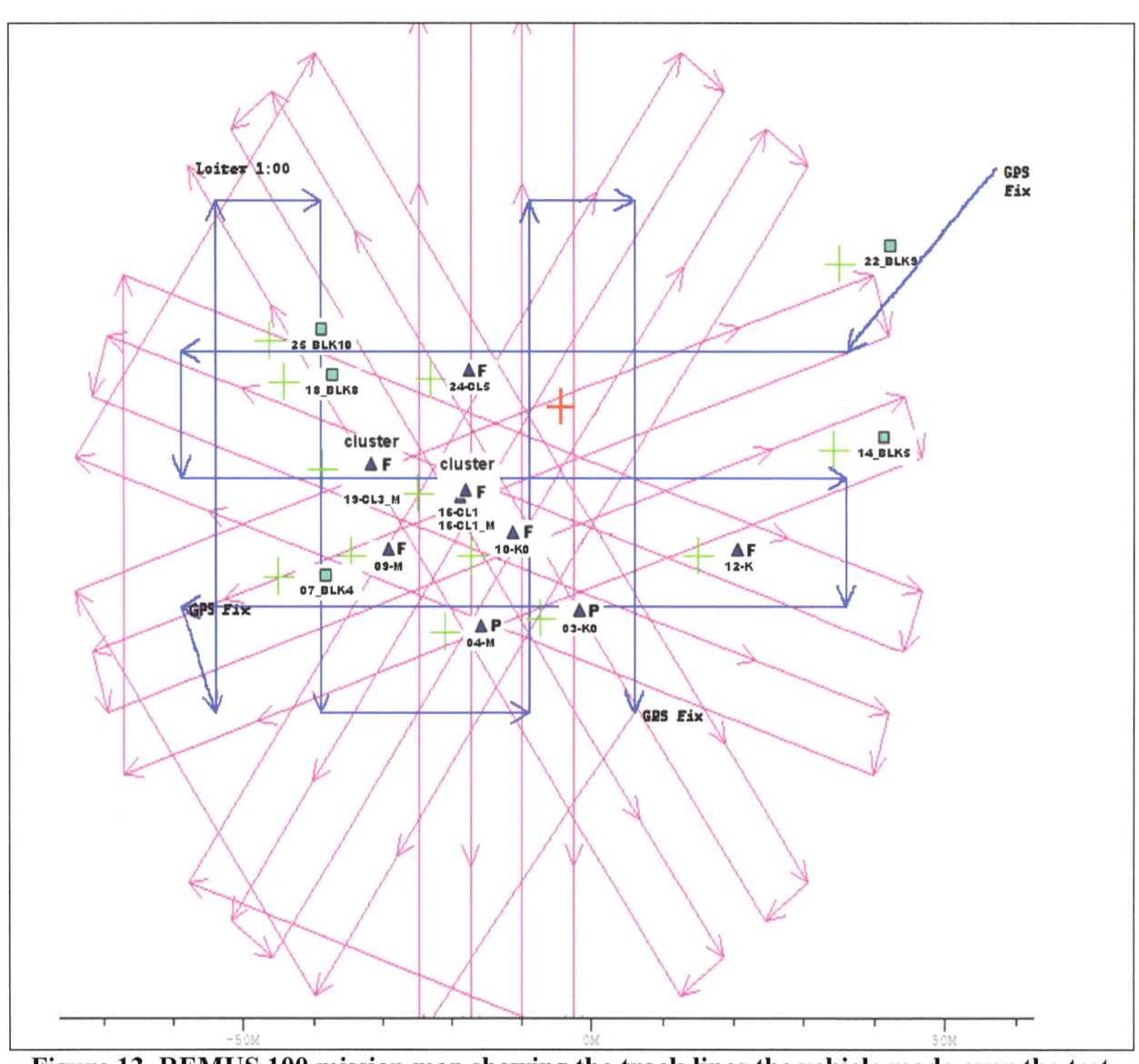

Figure 13. REMUS 100 mission map showing the track lines the vehicle made over the test field. The blue track lines are with the 900kHz sonar. The purple track lines are with the 1800 kHz sonar. The green (+) symbol represents the locations of the proud and flush buried UXO targets detected. The red (+) symbol represents the location of a UXO target relocated from Cluster 16-CL1_M by shrimper's nets.

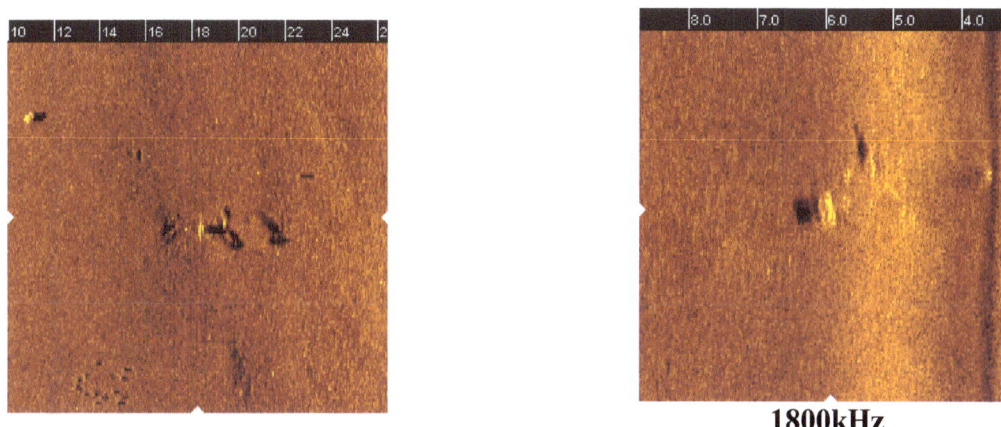

900kHz   **1800kHz**

Figure 14. 900 and 1800 kHz sonar imagery of 19-CL3_M

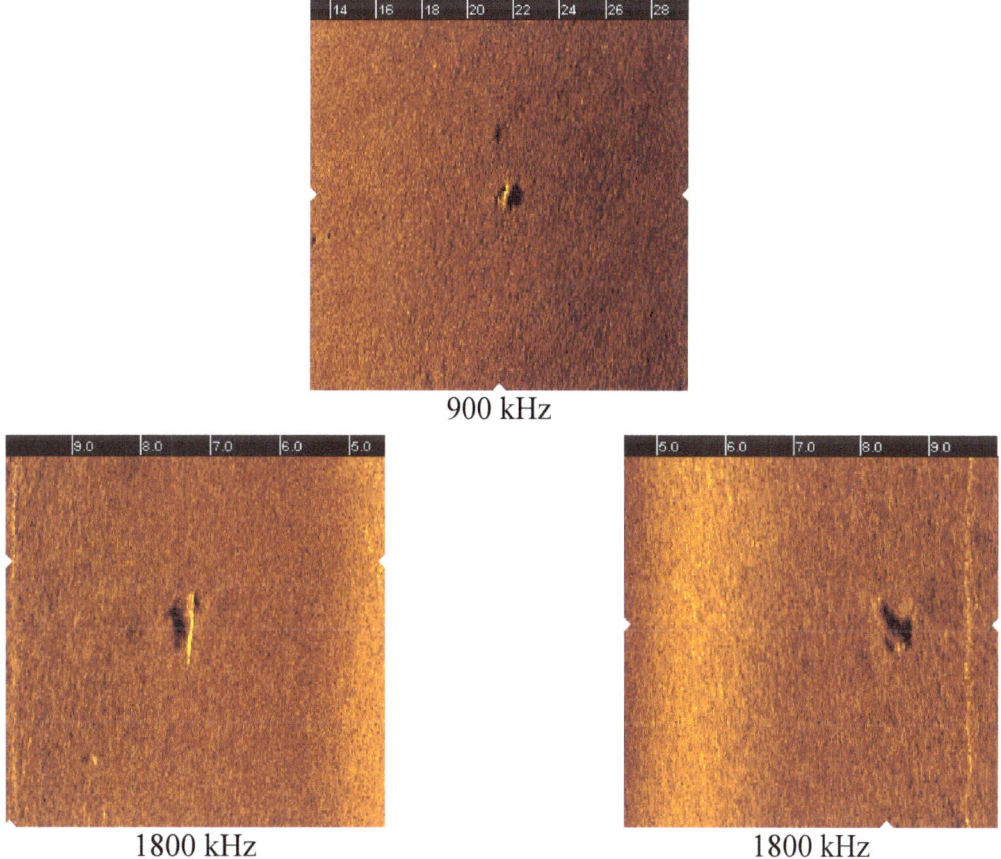

900 kHz

1800 kHz   1800 kHz

Figure 15. 900 and 1800 kHz sonar imagery of 09-M

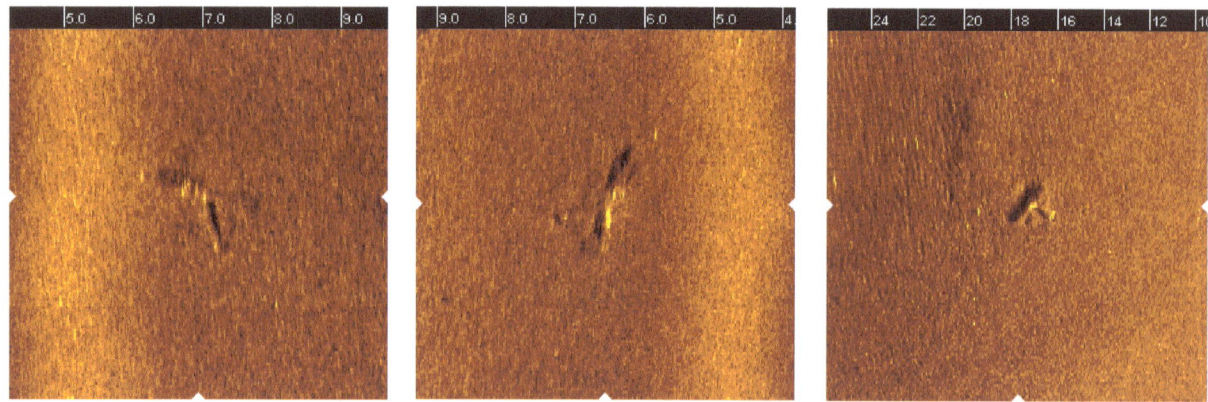

Figure 16. 1800 kHz sonar imagery of 24-CL5

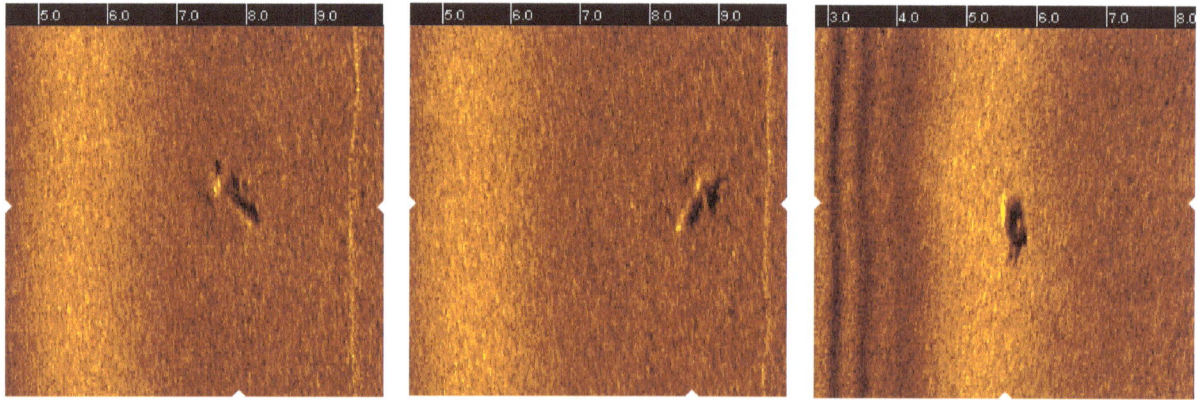

Figure 17. 1800 kHz sonar imagery of 04-M

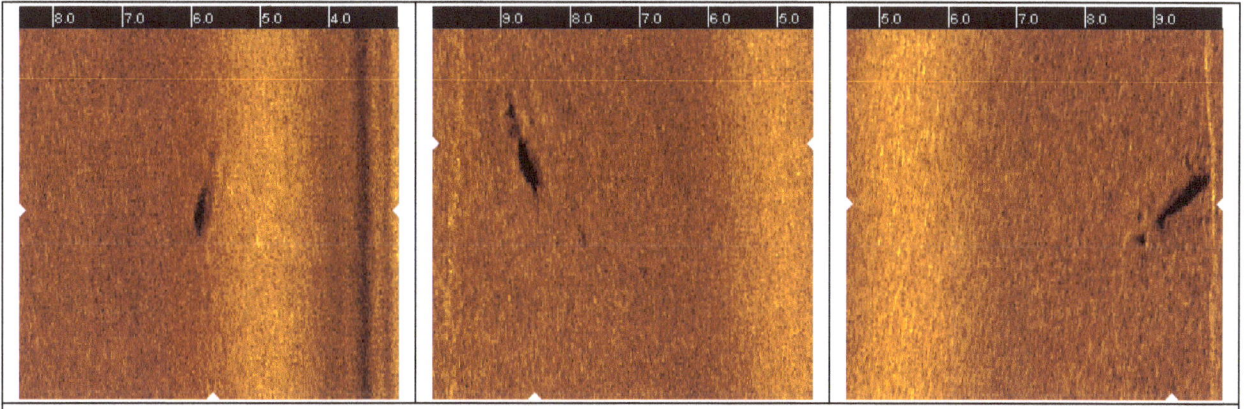

Figure 18. 1800 kHz sonar imagery of 10-K0

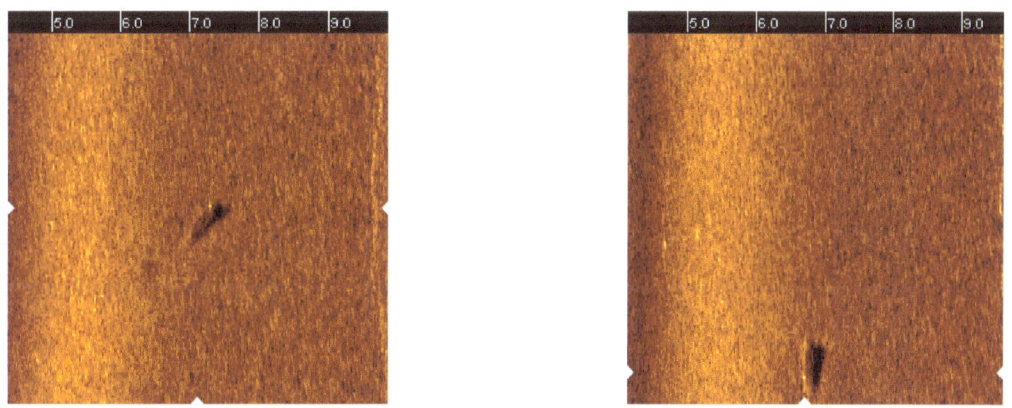

Figure 19. 1800 kHz sonar imagery of 16-CL1 and 16-CL1_M

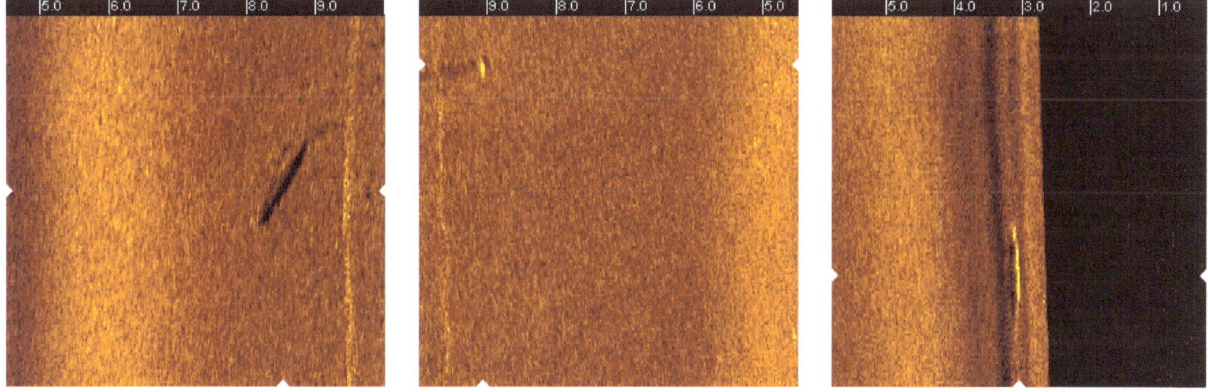

Figure 20. 1800 kHz sonar imagery of 03-K0

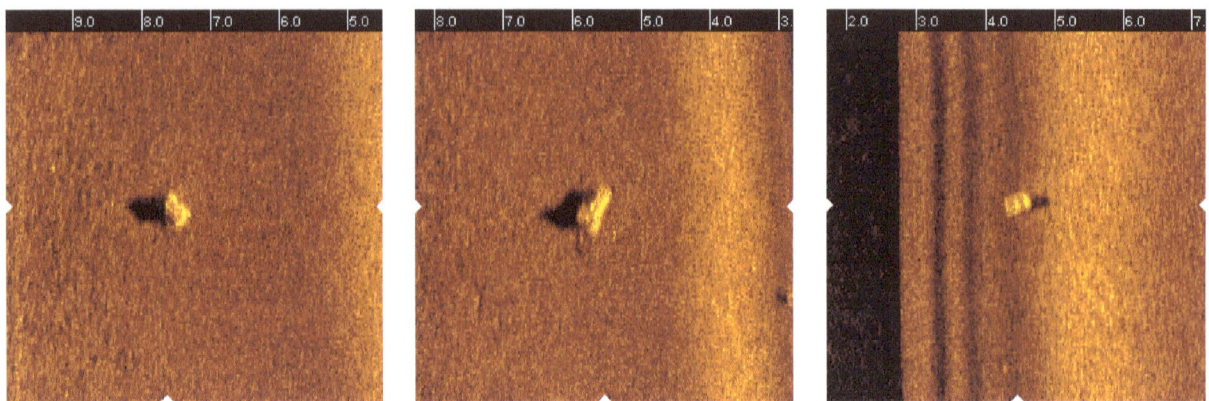

Figure 21. 1800 kHz sonar imagery of 07_BLK4

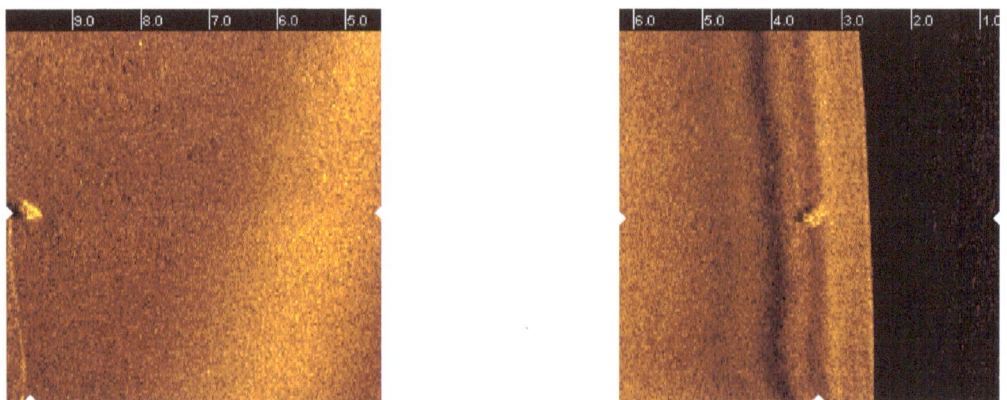

Figure 22. 1800 kHz sonar imagery of 22_BLK9

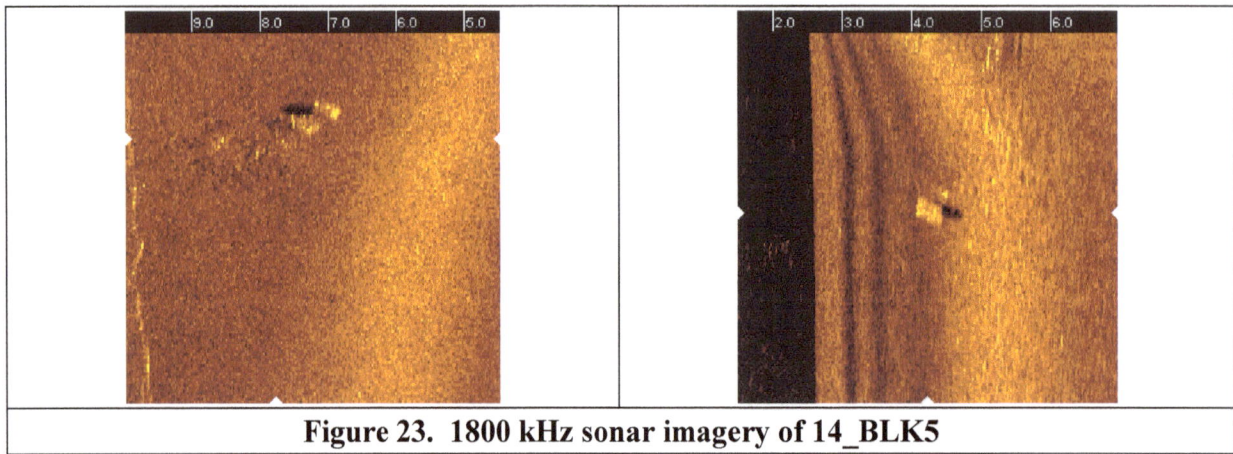

Figure 23. 1800 kHz sonar imagery of 14_BLK5

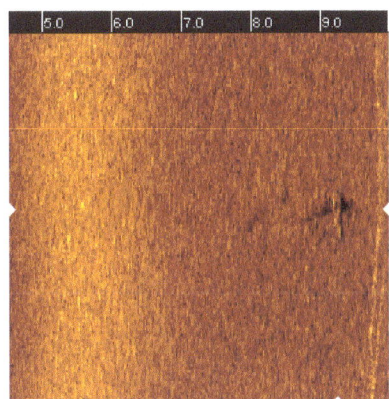
Figure 24. 1800 kHz sonar imagery of 12-K

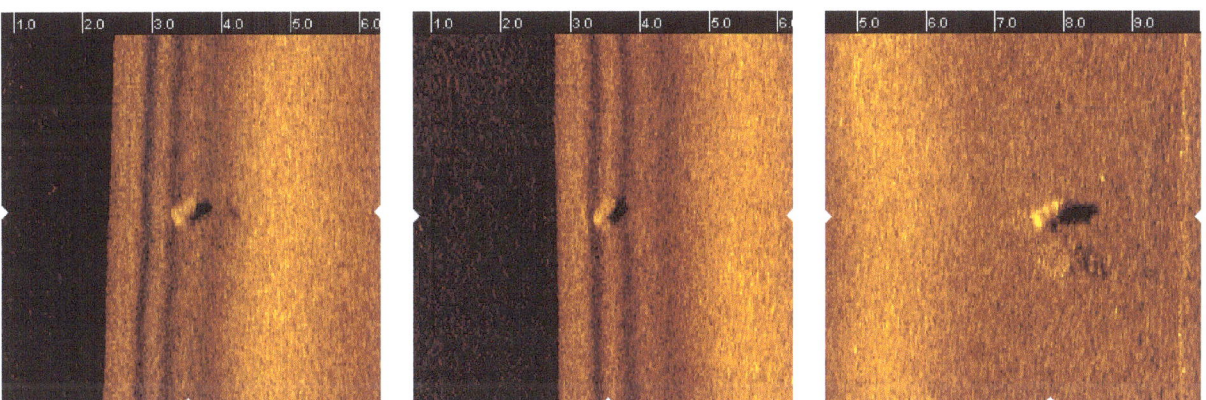
Figure 25. 1800 kHz sonar imagery of 25_BLK10

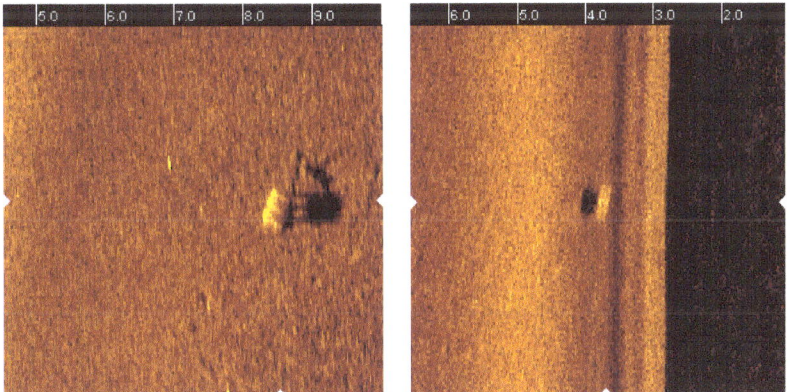
Figure 26. 1800 kHz sonar imagery of 18_BLK8

7.2 Buried Object Scanning Sonar (BOSS) Survey Lines

The Bluefin12 UUV performed seven survey missions over the ESTCP Site 4 target field. Two survey missions running in north-south/south-north and east-west/west-east headings at an altitude of 3 meters with survey line separations of 2 meters were performed on 25 and 26 June

2011. A survey mission running east-west/west-east headings at an altitude of 5 meters with survey line separations of 5 meters was performed on 27 June 2011. Survey missions over four targets at headings parallel to the orientations of the major-axis of the targets obtained from analysis of the previous survey missions were also performed on 27 June 2011. A summarized description of the surveys performed over the ESTCP Site 4 target field is provided in Table 4.

Table 4. Listing and description of the surveys over ESTCP Site 4 target field.

Survey Date	UUV Altitude	Line Separation	# of Survey Lines (heading)	Targets
June 25, 2011	3 meters	2 meters	25 lines (N-S/S-N) 25 lines (E-W/W-E)	
June 26, 2011	3 meters	2 meters	25 lines (N-S/S-N) 25 lines (E-W/W-E)	
June 27, 2011	5 meters	5 meters	15 lines (E-W/W-E)	
	3 meters	2 meters	10 lines (070-250 deg)	09-M
	3 meters	2 meters	10 lines (075-255 deg)	16-CL1
	3 meters	2 meters	10 lines (110-290 deg)	12-K

The navigation tracks of the survey missions can be displayed for referencing survey lines and target localization through the BOSS NavTrack display window. Figure 27 illustrates the BOSS NavTrack window with the ESTCP Site 4 target field in the center. The horizontal and vertical lines represent latitude (30:08.DDDD) and longitude (-85:41.DDDD), respectively.

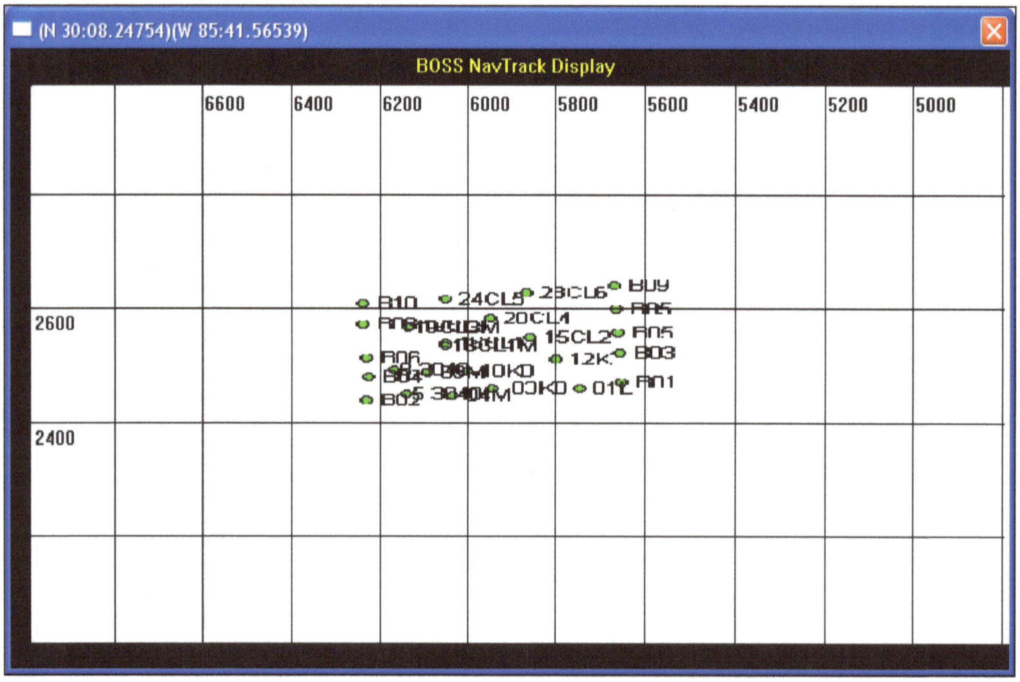

Figure 27. BOSS NavTrack display window that displays survey mission lines.

Figures 28 and 29 are representative of the navigation tracks for the surveys performed on 25 and 26 June 2011. Twenty-five tracks ran in the east-west/west-east heading and 25 tracks ran in the north-south/south north heading.

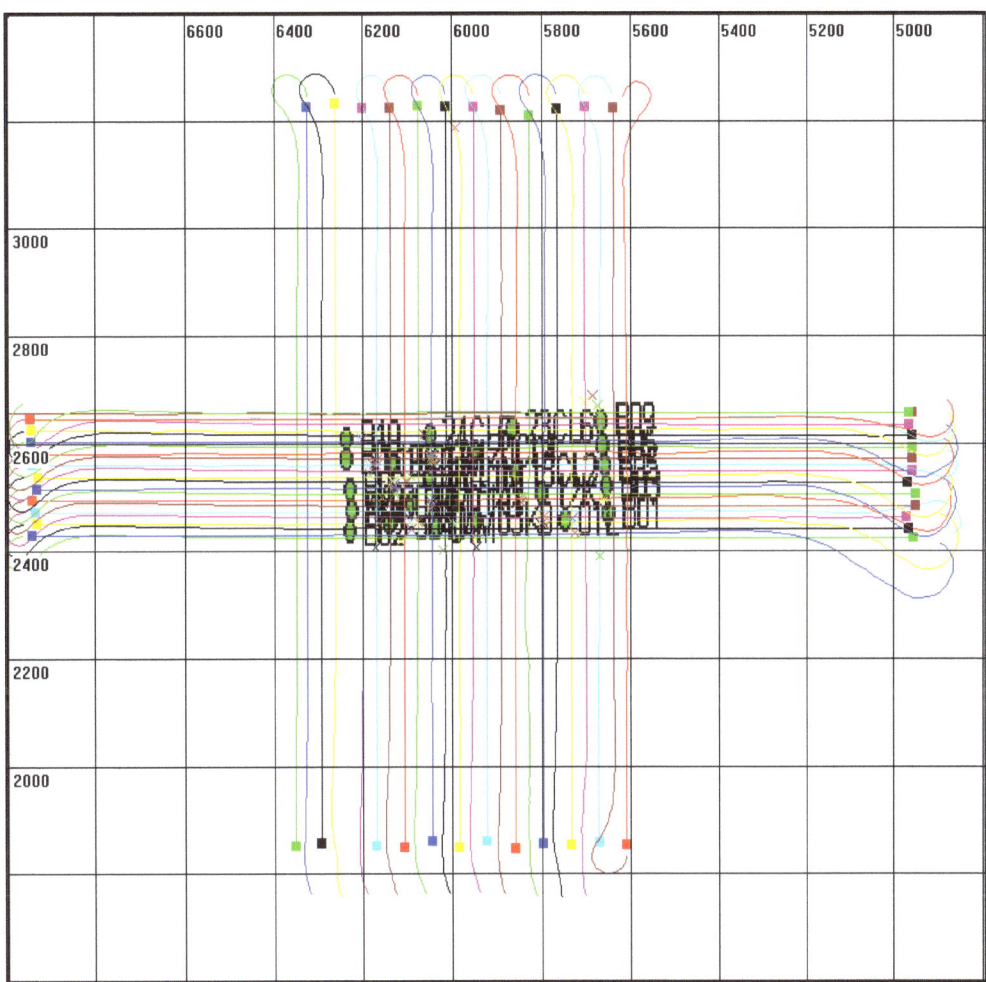

Figure 28. Navigation tracks for the survey missions performed on 25 June 2011.

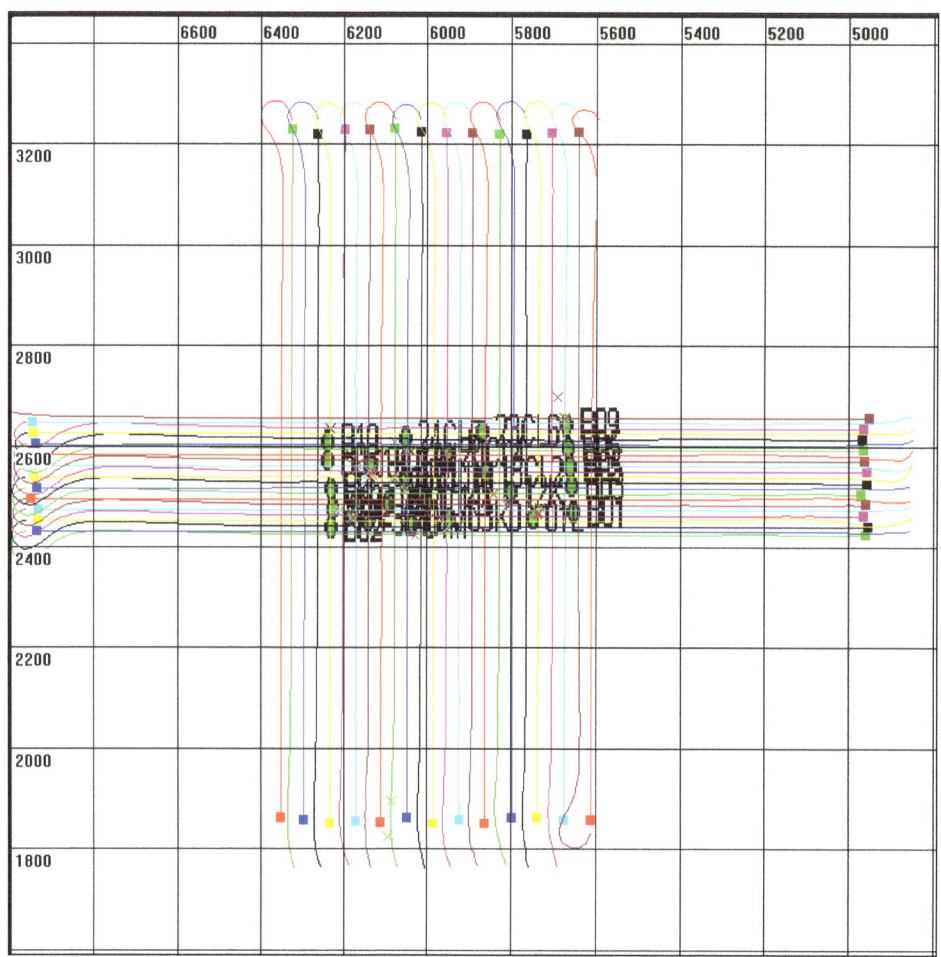

Figure 29. Navigation tracks for the survey missions performed on 26 June 2011.
Figure 30 is representative of the navigation tracks for the 5 meter altitude surveys performed on 27 June 2011. Fifteen tracks ran in the east-west/west-east heading. The higher altitude surveys were performed to show the wider field-of-view of the BOSS sonar.

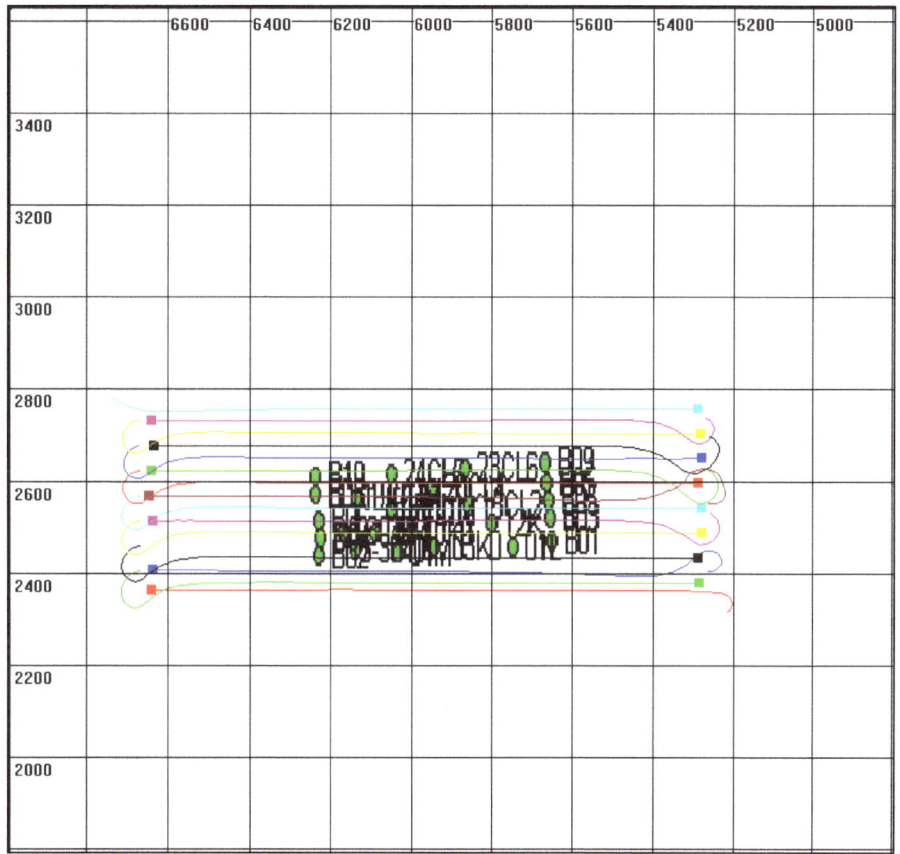

Figure 30. Navigation tracks for the survey missions performed on 27 June 2011.
Survey missions over selected targets at an altitude of 3 meters and at headings parallel to the major axis of the target were also performed on 27 June 2011. Figures 31 through 33 are representative of the navigation tracks for these missions over targets 09-M, 16-CL1, and 12-K, respectively.

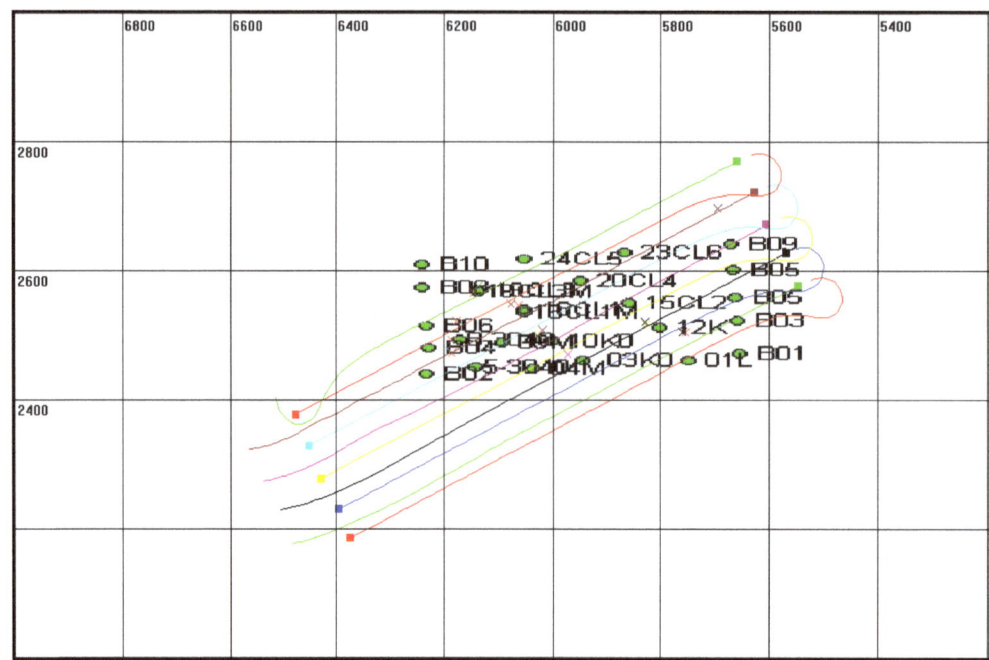

Figure 31. Navigation tracks for the survey missions over target 09-M.

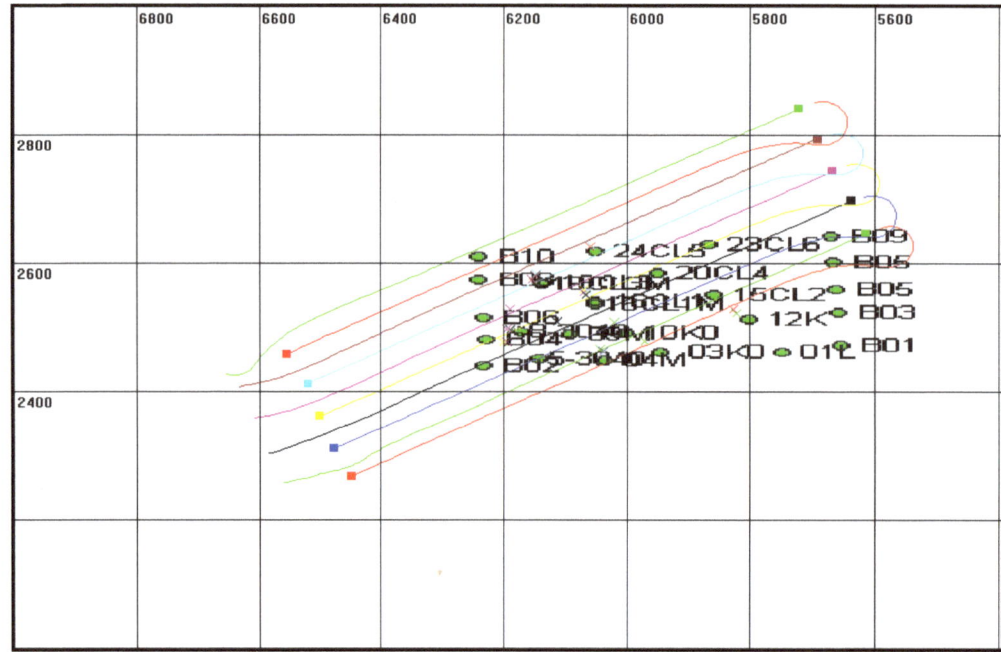

Figure 32. Navigation tracks for the survey missions over target 16-CL1.

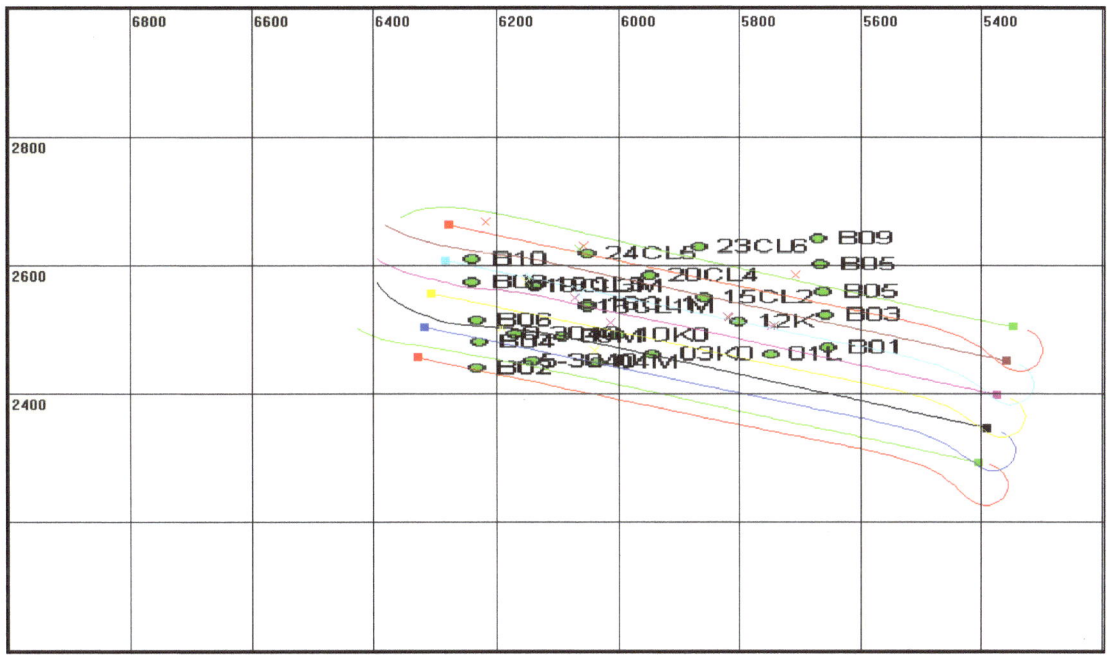

Figure 33. Navigation tracks for the survey missions over target 12-K.

7.3 Buried Object Scanning Sonar (BOSS) Survey Missions

The BOSS system detected most all M-type and all the larger UXO targets in both proud and buried states; however, detection of the clusters made up of the 50-caliber munitions was not accomplished with high confidence. Figures 34 through 40 show BOSS imagery from selected survey lines of the UXO targets listed in Table 3 of Section 4.1.1 that are highlighted in green.

The BOSS images consist of the X-Y plan view showing the plumbed position of the target represented by an "X" and the image of the target of interest highlighted with a circle and labeled with the target's designation. A zoomed, 3D multi-aspect BOSS image for each target is also provides the X-Z and Y-Z perspective views to show target burial state. BOSS target localization, distance from BOSS localization to diver plumbed positions ("X"), target dimensions and burial state information for each of the targets is provided from BOSS PMA. Note that all BOSS-generated images in this report are for cases with sub-critical grazing angles in the sand bottom field.

Figure 34 shows BOSS X-Y and zoom multi-aspect 3D imagery of 12-K. The BOSS localization for target 12-K is.41 meters north-east and 4.80 meters south of its plumbed position, respectively. The imagery also shows a cylindrical flushed buried target with approximate measurements in length and width of 1.33 and 0.27 meters, respectively.

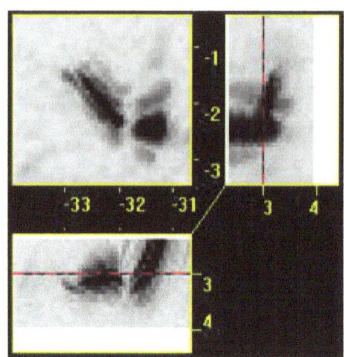

Figure 34. BOSS zoom imagery for target 12-K. Vertical axes are cross track, and horizontal axes are along track, in meters.

Figure 35 shows BOSS X-Y and zoom multi-aspect 3D imagery of targets 03-K (red circles) and 04-M (yellow circles). Figure 35a shows a snippet of the X-Y plan view for a snippet of survey line 20110626_001_Av000. Targets 03-K and 04-M are 4.08 and 3.63 meters south of the plumbed positions, respectively. Figure 35b is the BOSS zoomed image for target 04-M. The imagery shows the target in a proud status with approximate measurements in length and width of 0.78 and 0.18 meters, respectively. Figure 35c is the BOSS zoom image for target 03-K. The imagery shows a partially buried target with approximate measurements in length and width of 1.34 and 0.24 meters, respectively.

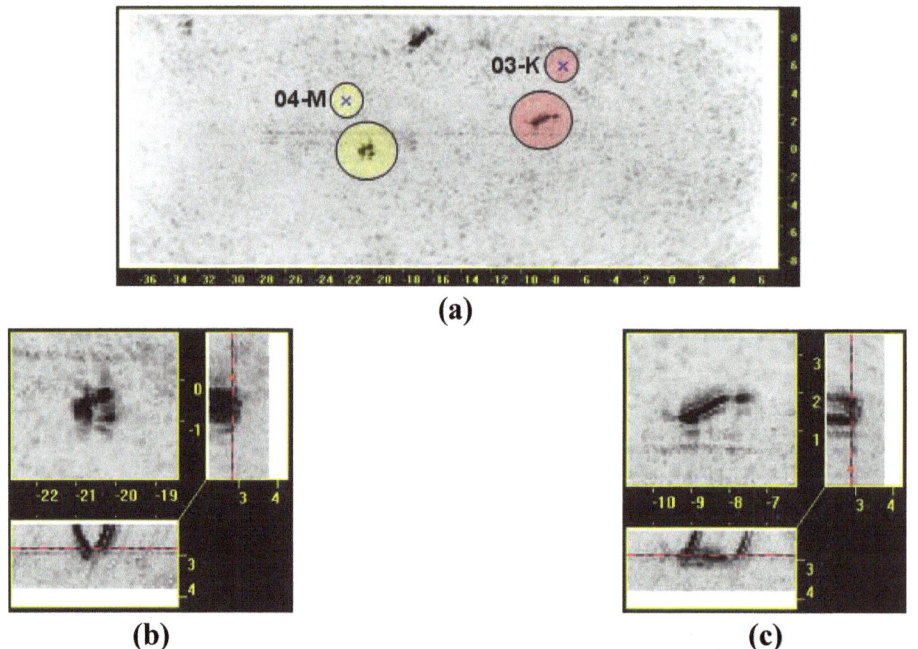

Figure 35. BOSS imagery for targets 04-M and 03-K: (a) plan view image, (b) zoom image for target 04-M and (c) zoom image for target 03-K. Vertical axes are cross track, and horizontal axes are along track, in meters.

Figure 36 shows BOSS X-Y and zoom imagery for target 10-K. Figure 36a shows the X-Y plan view for a snippet of survey line 20110626_001_Ap000. Target 10-K is 3.05 meters south of its plumbed position. Figure 36b is the BOSS zoom image for target 10-K. The imagery shows a proud target with approximate measurements in length and width of 1.30 and 0.27 meters, respectively.

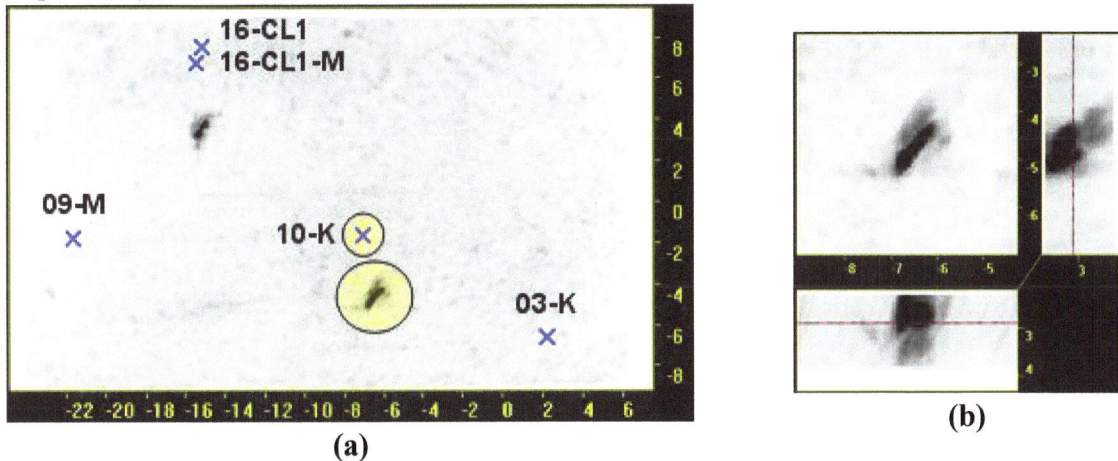

(a)

(b)

Figure 36. BOSS imagery for target 10-K: (a) plan view image and (b) zoom image of target 10-K. Vertical axes are cross track, and horizontal axes are along track, in meters.

Figure 37 shows BOSS X-Y and zoom imagery for target 09-M. Figure 37a shows the X-Y plan view for a snippet of survey line 20110626_002_Ap000. Target 09-M is 7.82 meters south-east of its plumbed position. Figure 37b is the BOSS zoomed image for target 09-M. The imagery shows a proud target with approximate measurements in length and width of 0.80 and 0.17 meters, respectively.

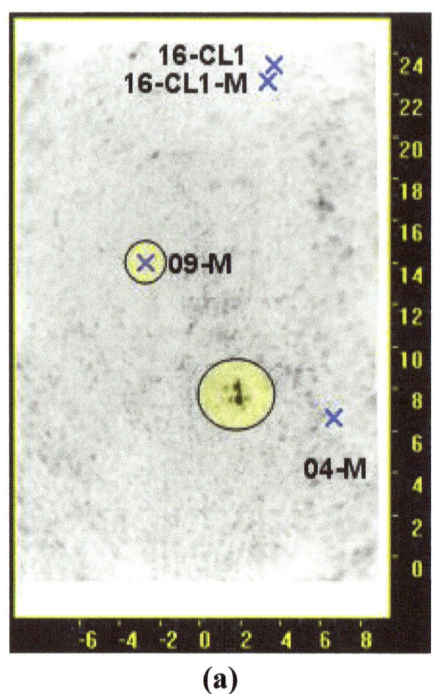

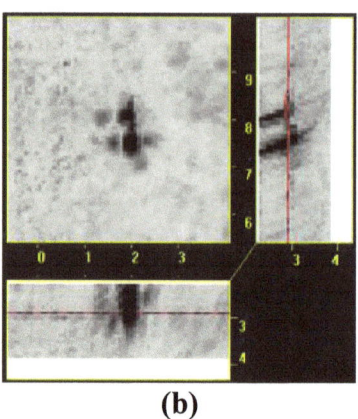

(b)

(a)

Figure 37. BOSS imagery for the target 09-M: (a) plan view image and (b) zoom image of target 09-M. Horizontal axes are cross track, and vertical axes are along track, in meters.

Figure 38 shows BOSS X-Y and zoom imagery for two targets 19-CL3-M. Figure 38a shows the X-Y plan view for a snippet of survey line 20110626_001_Aj000. The targets are 4.22 and 3.92 meters south of the plumbed position. Figure 38b is the BOSS zoom image for the two targets 19-CL3-M. In this survey line, the two small targets 19-CL3-M are clearly discerned. The image shows the two targets flushed buried with approximate measurements in lengths and widths of 0.70 and 0.17 meters, respectively.

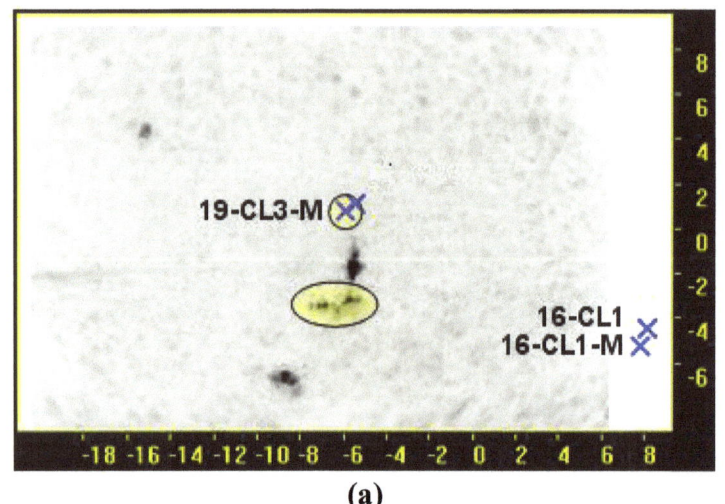

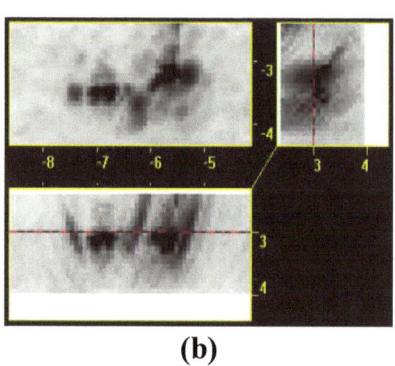

(b)

(a)

Figure 38. BOSS imagery for the targets 19-CL3-M: (a) plan view image and (b) Zoom image for targets 19-CL3-M.

Figure 39 shows BOSS X-Y and zoom imagery for target 16-CL1. Figure 39a shows the X-Y plan view for a snippet of survey line 20110626_001_Am000. The target is 3.75 meters south of its plumbed position. Figure 39b is the BOSS zoom image for target 16-CL1. The image shows a proud target with approximate measurements in length and width of 1.38 and 0.27 meters, respectively. Similarly to target 19-CL3-M, two small targets were placed in close proximity to a larger target. For this case, only one of the small targets could be discerned just off of the end of the larger target. The other smaller target could not be discerned and most likely was placed right next to the larger target.

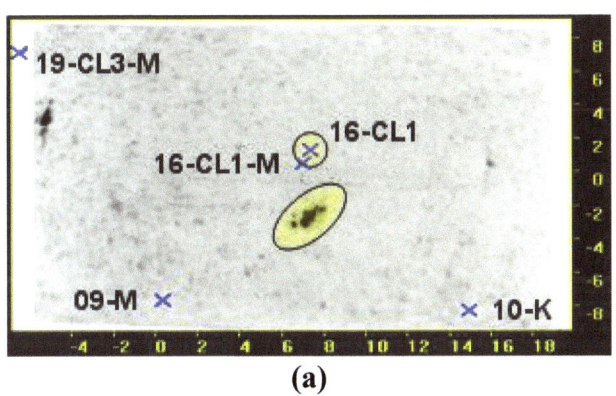

(a)

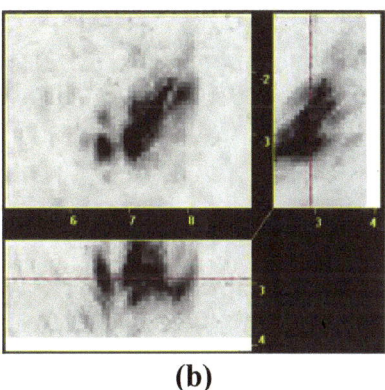

(b)

Figure 39. BOSS imagery for the target 16-CL1: (a) plan view image (b) Zoom image of target 16-CL1.

Figure 40 shows BOSS X-Y and zoom imagery for target 24-CL5 that consists of a cluster of four small targets. Figure 40a shows the X-Y plan view for a snippet of survey line 20110626_001_Af000. The image shows a cluster of targets 4.20 meters south of their plumbed position. Figure 40b is the BOSS zoom image shows a cluster of targets with mean approximate measurements in length and width of 0.65 and 0.21, respectively.

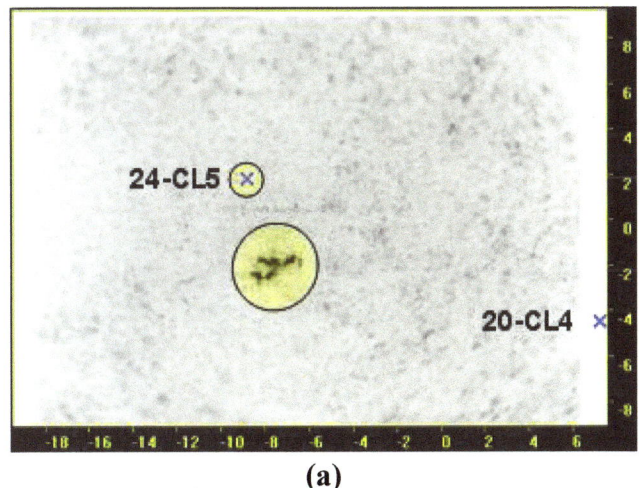

(a)

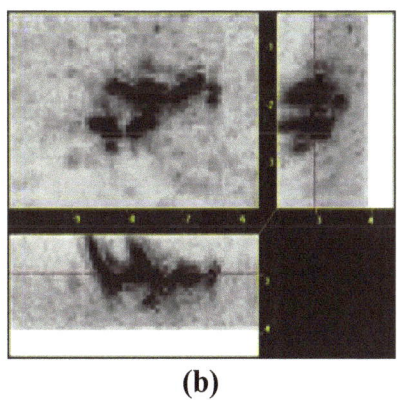

(b)

Figure 40. BOSS imagery for the target 24-CL5: (a) plan view image and (b) Zoom image of target 24-CL5.

Table 5 provides a listing of the targets, BOSS localizations and ranges to plumbed positions, burial status and target dimensions obtained via BOSS PMA. In summarizing, several survey missions were performed over the target field with the primary survey lines being performed with separation of 2 meters and UUV altitude of 3 meters in north-south and east-west headings. These survey lines, as shown in Figures 28 and 29, provided coverage of the entire target field and the BOSS system was able to detect 13 of the 14 targets that were deployed. One of the two targets in 16-CL1-M cluster was not clearly discerned. The average offset between the BOSS target localization and the plumbed position was approximately 3.8 meters; however, target 09-M had an offset distance of 7.8 meters from its plumbed position.

The burial state from BOSS PMA was in agreement with the target deployment plan for five targets and these are color coded in green in the Target Status column in Table 5. The burial states from BOSS PMA for nine targets were not in agreement with the target deployment plan and these are color coded in red. This does not indicate that the BOSS system was incorrect; for example, target 10-K according to the deployment plan was supposed to be flushed buried but in the BOSS PMA the target is proud. This will be discussed in detail in the following section.

From Section 4.1.1, the target dimensions are the following: Target type K has length and diameter of 1.64 and 0.27 meters, respectively and target type M has length and diameter of 0.68 and 0.16 meters, respectively. The mean length and width extracted from BOSS measurements for the K target types (12-K, 03-K, 10-K and 16-CL1) is 1.34 and 0.26 meters, respectively. The mean length and width extracted from BOSS measurements for the M target types (04-M, 09-M, 19-CL3-M, and 24-CL5) is 0.67 and 0.18 meters, respectively.

Table 5. Target listing with BOSS localization and target status information.

Target	BOSS Localization Longitude/Latitude	Range to Ground Truth (meters)	Target Status (burial state, length, width)
12-K	-85 41.580 / 30 08.249	4.80	flush buried, 1.33, 0.27
03-K	-85 41.595 / 30 08.244	4.08	partially buried, 1.34, 0.24
04-M	-85 41.603 / 30 08.243	3.63	proud, 0.78, 0.18
10-K	-85 41.600 / 30 08.247	3.05	proud, 1.30, 0.27
09-M	-85 41.606 / 30 08.246	7.82	proud, 0.80, 0.17
19-CL3-M	-85 41.614 / 30 08.254	4.22	flush buried, 0.68, 0.17
	-85 41.613 / 30 08.254	3.92	flush buried, 0.70, 0.17
16-CL1 (K)	-85 41.605 / 30 08.252	3.75	partially buried, 1.38, 0.27
16-CL1-M 1	-85 41.606 / 30 08.251	3.70	partially buried, 0.61, 0.15
2	Not Detected		
24-CL5 (M) 1	-85 41.605 / 30 08.260		partially buried, 0.65, 0.17
2	-85 41.605 / 30 08.259	4.20	partially buried, 0.58, 0.21
3	-85 41.604 / 30 08.260		partially buried, 0.54, 0.21
4	-85 41.603 / 30 08.260		partially buried, 0.64, 0.21

One of the data products provided by PMA analysis of the BOSS sonar imagery; specifically for cylindrical targets is target orientation. This can be of upmost importance during recovery operations of flush to deep buried targets where orientation of the target is desired for proper rigging and handling. Table 6 lists the target orientations obtained from PMA analysis of the BOSS sonar imagery from the surveys performed on 26 June and shown in Figures 34, 37 and 39 for targets 12-K, 09-M and 16-CL1, respectively.

Table 6. BOSS orientation results obtained from low and high altitude surveys.

TARGETS	BOSS Target Heading
12-K	110-290 deg
09-M	070-250 deg
16-CL1	075-255 deg

7.4 PERFORMANCE OBJECTIVES – Buried Object Scanning Sonar (BOSS)

The performance objectives for BOSS system under the Demonstration Plan are summarized in Table 7. The objectives include demonstration of the performance of the probability of detection, determine the degree of burial size, shape, and orientation of contacts and target localization accuracy against proud and fully buried UXO targets.

Table 7. Performance Objectives for BOSS Sensor

Performance Objective	Metric	Data Required	Success Criteria
Quantitative Performance Objectives			
Detection of all munitions of interest	Percent detected of seeded items	• Location of seeded Items	Pd=0.90
Determine the degree of burial, size, shape, and orientation of contacts.	Percent correct classification of the burial depth, size, shape, and orientation of the seeded targets.	• Validation data for selected targets	Demonstration of >90% correct classification of the size, shape, and orientation of the targets. Length and burial measurements considered correct shall be within 25% of actual.
Location accuracy	Average localization error in meters from latitude and longitude ground truth localization for seed items	• Location of seed items surveyed to accuracy of 1m	ΔLocalization Error < 1m from ground truth
Qualitative Performance Objectives			
Ease of use		• Feedback from technician on usability of technology and time required	

7.4.1 Objective: Detection of All Munitions of Interest - BOSS

This objective focuses on the effectiveness of the BOSS sensor in detecting proud to full buried small UXO targets and discriminating targets that are clustered together.

7.4.1.1 - Metric

A total of fourteen UXO targets, ten class M targets and four class K targets were deployed in the field. Figure 12 illustrates the layout of the target field and Table 3 provides a listing of the target's plumbed position and description.

7.4.1.2 - Data Requirements

BOSS data analysis demonstrated that of the fourteen targets seeded in the field, thirteen were detected; hence, 93% of the targets were detected.

7.4.1.3 - Success Criteria

The smaller 50-caliber shells were deployed in the field as extra targets to assess the detection and imaging capability of higher frequency imaging sonar systems that were operated over the same area. These target items are not part of the target list used to assess the Bluefin12 UUV capability of detecting underwater UXO targets. Presently, the BOSS sonar cannot resolve these small targets.

The BOSS detected and localized thirteen of fourteen seeded targets with high operator confidence resulting in a probability of detection of 0.93. The objective to detect the targets with a probability of detection of 0.90 or above was met.

7.4.2 Objective: Determine the degree of burial, size, shape, and orientation of contacts

This objective focuses on determining the effectiveness of the BOSS sensor in the capability of determining the degree of burial, size, shape, and orientation of the UXO target types with high confidence.

7.4.2.1 – Metric

The metric used to asses this objective is to identify the number of targets correctly correlated with the target type seeded at the plumbed position.

7.4.2.2 – Data Requirements

High confidence in discriminating the three classes of targets in size and shape is clearly shown with BOSS imaging and the extraction of target physical dimensions. Two types of cylindrical-shaped UXO targets were seeded: class M targets with typical length and width of 0.68 and 0.16 meters, respectively, class K targets with typical length and width of 1.64 and 0.27 meters, respectively.

The ability in determining the orientation on the parallel heading of the major axis of the targets and the degree of burial of the targets was lost, due to, not being able to recover the field after the test. With this delay in recovering the field and the effects of natural (tides, weather) and manmade (shrimpers dragging nets) events occurring to change target placement, the degree of burial at this time would not be represented to the degree of burial at the time of the test.

7.4.2.3 – Success Criteria

Target measurements obtained from BOSS imagery were consistent with the actual physical dimensions of each of the class of targets detected at or near the plumed positions. For the class K targets, the average length of the target measured from the image was 1.34 meters as compared to 1.64 meters and for the width it was 0.26 meters as compared to 0.27 meters. For the class M targets, the average length of the target measured from the image was 0.67 meters as compared to 0.68 meters and for the width it was 0.18 meters as compared to 0.16 meters. Although the measurements are not exact, the BOSS imagery clearly discriminates the different classes of targets. The BOSS measured the seeded targets with high operator confidence resulting in a probability of correct size measurement within 18% of actual size. The objective to measure the targets with a probability of within 25% of actual size was met.

The BOSS sonar PMA results for target orientation of the parallel headings of the major axis couldn't be verified as being correct with a probability of greater than 90%. The demonstration of >90% correct identification of the orientation of the targets was not confirmed due in part not being able to recover the field after the test..

7.4.3 Objective: Location Accuracy - BOSS

This objective focuses on the effectiveness of the BOSS sensor for localization of the UXO targets.

7.4.3.1 – Metric

The metric to assess this objective is to compare the BOSS target localization with the diver provided plumbed positions.

7.4.3.2 – Data Requirements

BOSS data was collected over the seeded field and analysis followed to detect and localize the targets of interest. The BOSS localization for each target was cataloged and verified against the target listing and the plumbed positions provided by the divers.

7.4.3.3 – Success Criteria

The area for seeding the UXO targets was selected due to it being clean from debris or clutter. As stated and shown earlier, the BOSS detected 15 of 16 deployed targets with high confidence. The results have been shown in image form in Figures 35 through 41 and target locations both the plumbed positions and those extracted from BOSS PMA are listed in Tables 3 and 5. The average offset between the plumbed position and the BOSS localization was approximately 3.8 meters. However, target 09-M had an offset of approximately 7.8 meters from its plumbed position; this offset most likely occurred during deployment. Analysis of the localization offset shows a bias towards the south when referenced to the target plumbed position.

In order to consider this objective a success, the criteria of localizing 95% of the detected targets to be within the 1 meter from the diver reported ground truth localizations had to be achieved. The results show that this objective was not met; however, two sources of error can have an impact on these controlled tests. The first source for introducing error is the method of obtaining the plumbed position (ground truth). Typically, a line is tied to the target and once the target is deployed, the line is pulled tight up to the surface and a GPS reading at that position is obtained.

A problem with this method is that the surface craft will drift and the actual GPS reading may not be directly over the target; this error will also increase in deeper waters. The second source is the present navigation package in the Bluefin12 UUV. The Bluefin12 UUV navigation package consists of a compass, an Inertial Measurement Unit (IMU) and a Doppler velocity log (DVL). This package provides navigation accuracy to within 0.5% of distance traveled. Prior to the exercise, compass calibrations must be performed to calibrate the navigation solutions and this was performed at the beginning of the survey missions. Thus, the primary source of error in target localization will be in the accuracy of the GPS reading on the plumbed position.

Another UUV system integrated with the Laser Scalar Gradiometer (LSG), a UUV-based magnetic sensor with very high sensitivity that has been successfully demonstrated for mine localization was deployed and surveyed the UXO seeded field on 16 June 2011. Localization results from the LSG and BOSS sensors are plotted with the plumbed positions and shown in Figure 41. The red diamonds, blue squares and green triangles are representative of the plumbed positions, LSG, BOSS localizations, respectively. In this graph, LSG localizations for all survey lines are plotted. The graph also shows that for most all targets, the LSG and the BOSS localizations are consistent and the LSG localizations also show offsets similar to the BOSS with respect to the plumbed positions. This clearly shows the problem in the method employed to ground truth underwater targets.

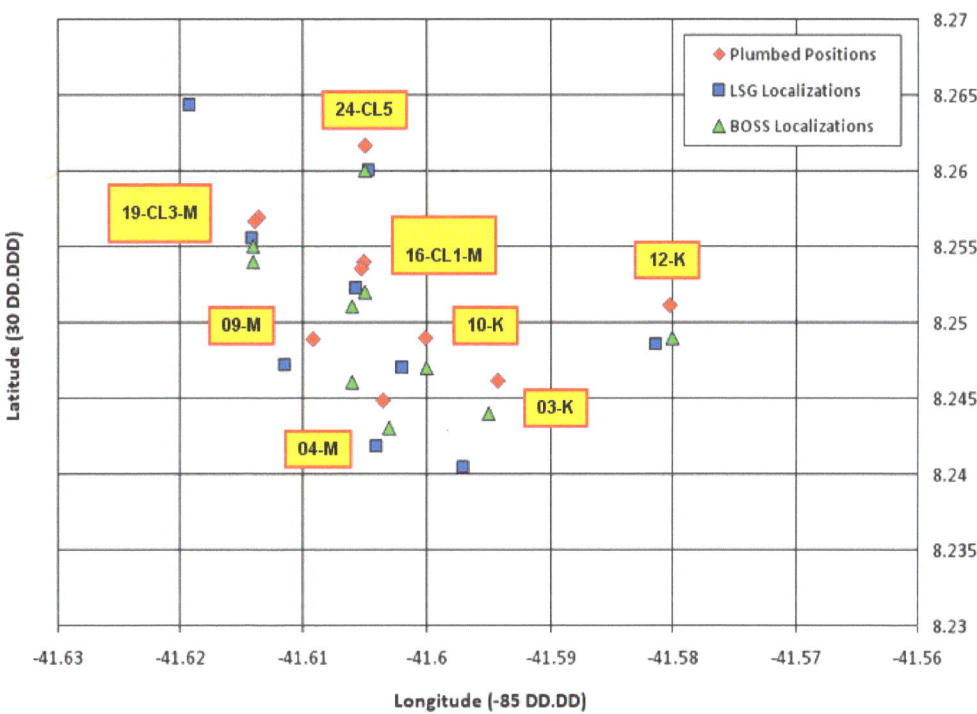

Figure 41. BOSS and LSG target localizations.

7.4.4 Objective: Ease of Use - BOSS

This objective provides operator feedback on the ease of use of the BOSS and its data PMA efforts.

7.4.4.1 – Data Requirements

The installation of the BOSS primary components (receiver wings, transmitter and electronics bottle) onto the Bluefin12 UUV is straightforward and takes approximately one hour to complete. Pre-mission testing usually takes a couple of hours which includes deploying the UUV in the water for ballasting the UUV to the right buoyancy, checking for UUV and sensor for ground faults and conducting a check of the BOSS sensors in water.

The UUV operator programs the sensor to operate at different states (power on and record, power on and standby, power off) when planning the survey missions. The BOSS operator can query the UUV via FreeWave radio to obtain BOSS sensor status which includes operation and data recording status when the UUV surfaces after a survey run. However, data quality can only be confirmed during PMA.

The BOSS PMA effort is also a straightforward two-step process: 1) upon recovery of the UUV after mission completion, all data is downloaded via an Ethernet connection; 2) The data analyst will set up the BOSS PMA application to automatically sequence through the data files and generate (X-Y) plan view images for each survey line. These images provide the first baseline analysis of the BOSS data by showing target detections on each of the survey lines. Further, in-depth analysis follows by selecting survey lines that show the targets of interest and re-processing these to generate multi-aspect (X-Y, X-Z and Y-Z) 3D target images and extract target information.

7.5 Summary – Buried Object Scanning Sonar (BOSS)

The BOSS sensor clearly demonstrated the capability of detecting and imaging UXO targets of the class M targets in proud, partially buried and fully buried configurations. Discriminating between the different targets was accomplished by comparing the known target physical dimensions with the measured dimensions extracted from the BOSS imagery. However, discriminating multiple targets in a clustered configuration was difficult for the case of targets closely spaced together (target 16-CL1-M). This is due to the medium resolution of the BOSS sensor. In previous years, the developer of the BOSS system has recommended increasing the transmit frequency to increase the resolution to enhance the imagery but funding this effort has not been forthcoming.

As to target localization, the BOSS system receives and uses UUV navigation information to calculate target localization. The present navigation package in the Bluefin12 UUV can be updated to improve its accuracy. In addition, the present method of obtaining the ground truth of targets deployed underwater described in Section 7.4.3.3 can introduce errors especially in deeper water making it difficult to accurately assess target localization.

7.6 Real-time Tracking Gradiometer (RTG) System

The RTG is a small passive magnetic sensor using fluxgate magnetometers measuring 3-orthogonal magnetic-field vector components at 3 spatially separated points in space. The RTG sensor is encased in a watertight housing and is integrated onto the nose section of the Bluefin12 UUV. The data collected from the magnetometer channels is used to calculate the five independent tensor gradient components of the magnetic field which are processed to extract the magnetic moments and three-dimensional position of a ferrous target. As shown in Figure 42 the RTG magnetic sensor system is mounted in the nose of the Bluefin12 vehicle, right behind the nose cover, so as to locate the sensors as far from vehicle induced sources of magnetic noise as possible.

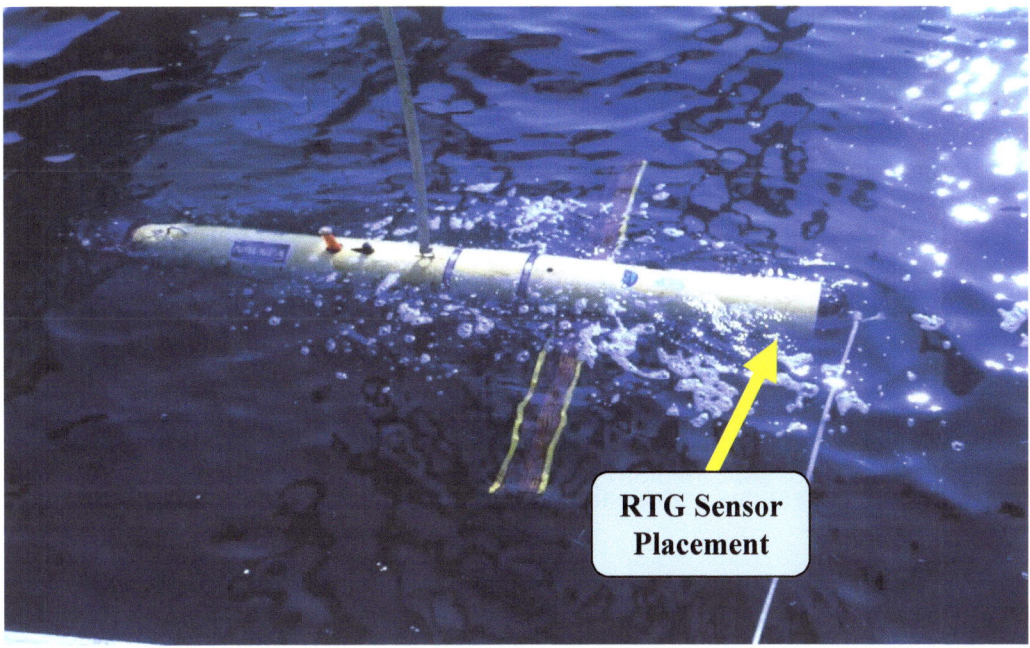

Figure 42. Bluefin 12 UUV on the surface of the water with the RTG sensor position indicated.

The data from 9 data channels (X, Y and Z for sensors A, B and C) is downloaded upon vehicle recovery along with the navigation data for post processing. The magnetic data are down sampled to correspond to the navigation data and then motion compensated in order to reduce the effects of the sensors moving through the Earth's magnetic field, which is otherwise seen as a change in signal. Five independent gradients are constructed from this sensor data and then magnetic dipoles are fit to segments of the signal which, when combined with the navigation data, allows the algorithm to localize ferrous targets.

7.6.1 ESTCP Site 4: Location, Target Field Layout and Survey Lines

Reference sections "4.0 SITE DESCRIPTION" for detailed descriptions of the site, targets and surveys.

7.6.2 RTG Survey Results

The RTG experienced hardware failure during the data collection event at the end of June, leading to lost data. A likely candidate for this failure is the oil-filled cables that connect the sensor head to the electronics bottle. These cables and their connection points have failed in the past with similar effects on the data. A refurbishment and recalibration of the RTG system would be beneficial for future testing events and will be necessary to improve results.

One of the data channels experienced intermittent failure allowing it to be used to construct a gradient for some of the runs. For times when this channel was faulty, a similar gradient could be constructed that would allow the data to be analyzed, albeit at a loss of accuracy and confidence. Unfortunately, a different data channel was completely broken for the entirety of the event and the nature of the gradient constructed using this channel made it impossible to construct a similar replacement gradient. Thus, of the 5 gradients to be used for target localization, only 4 were available and of those 4, only 3 were ideal at all times. Additionally, the collected data was fairly noisy and the signal-to-noise ratio was often very small for these sorts of small dipole targets, making data analysis difficult and the final results poor.

Tables 8 and 9 list the targets detected by the RTG and LSG, respectively with location, dipole strength, and orientation, distance offset from the vehicle and the bottom as well as the confidence number. Confidence number in this case refers to a number generated during the analysis that measures the quality of the fit of data to the optimal dipole model. For target signals that fit the magnetic dipole model well values above 9.0 are expected, whereas for a lower target signal strength it is appropriate to drop to a lower limit of 8.0 in order to insure all detections of interest have been found. Figure 43 shows LSG target locations with high and medium confidence localizations and localization of RTG targets compared to LSG and BOSS localizations.

Table 8. Target detections via the RTG System. "Y offset" is the horizontal distance to the target from the vehicle's track line, "Z offset" is the vertical distance to the target below the vehicle.

Target Latitude	Target Longitude	Dipole (A-m²)	Unit Vec (north)	Unit Vec (east)	Unit Vec (down)	Y Offset (m)	Z Offset (m)	Conf. Number
30 08.260	-85 41.594	64	-0.69	0.18	-0.71	0.77	-4.2	7.06
30 08.250	-85 41.605	23	0.73	0.47	0.50	-0.13	-3.1	9.13
30 08.245	-85 41.573	20	0.12	-0.69	0.71	0.41	-3.0	9.11
30 08.251	-85 41.584	14	0.60	-0.77	-0.21	2.1	-2.5	9.03
30 08.254	-85 41.610	33	0.77	0.61	0.19	0.58	-3.2	9.49
30 08.256	-85 41.619	80	0.97	0.25	0.07	-3.9	-2.6	9.52
30 08.256	-85 41.613	51	0.98	0.10	-0.17	0.43	-3.1	8.66
30 08.257	-85 41.617	34	0.93	0.37	0.06	0.14	-3.1	9.27
30 08.250	-85 41.602	52	0.76	-0.26	0.60	0.37	-4.0	8.59
30 08.265	-85 41.606	39	0.22	-0.04	0.97	-4.3	-1.4	9.97
30 08.255	-85 41.606	26	0.64	0.70	0.31	1.2	-3.0	9.54
30 08.251	-85 41.581	18	-0.98	-0.21	-0.07	-1.2	-2.9	8.34

Table 9. Target detections via the LSG System. The "Y off." is the distance to the target from the vehicle, "Z off." is the burial depth of the target.

Target Latitude	Target Longitude	Dipole (Am²)	Unit Vec (north)	Unit Vec (east)	Unit Vec (down)	Y Off. (m)	Z Off. (m)	Conf. #
30 8.246820	-85 41.575	38	0.31	-0.74	0.60	1.9	-2.5	9.90
30 8.247180	-85 41.6114	3.6	0.75	-0.34	0.56	-0.5	-2.1	9.89
30 8.243220	-85 41.5968	10	0.52	0.37	0.78	-1.1	-2.5	9.73
30 8.255520	-85 41.6141	62	0.9	0.15	0.41	1.7	-2.4	9.84
30 8.248500	-85 41.5814	24	0.77	-0.54	0.34	1.3	-2.1	9.84
30 8.252220	-85 41.6057	28	0.82	0.40	0.41	-1.7	-2.3	9.80
30 8.241780	-85 41.604	2.3	0.97	0.12	-0.20	-2.5	-1.7	9.78
30 8.247000	-85 41.602	7.1	0.66	0.21	0.72	2.0	-2.2	9.40
30 8.260020	-85 41.6046	5.3	0.95	-0.25	0.17	-4.0	-2.3	8.02
30 8.249220	-85 41.5959	1.1	-0.19	0.85	0.49	-0.2	-2.0	9.96
30 8.240400	-85 41.5971	13	-0.81	0.58	0.03	-0.1	-2.3	9.85

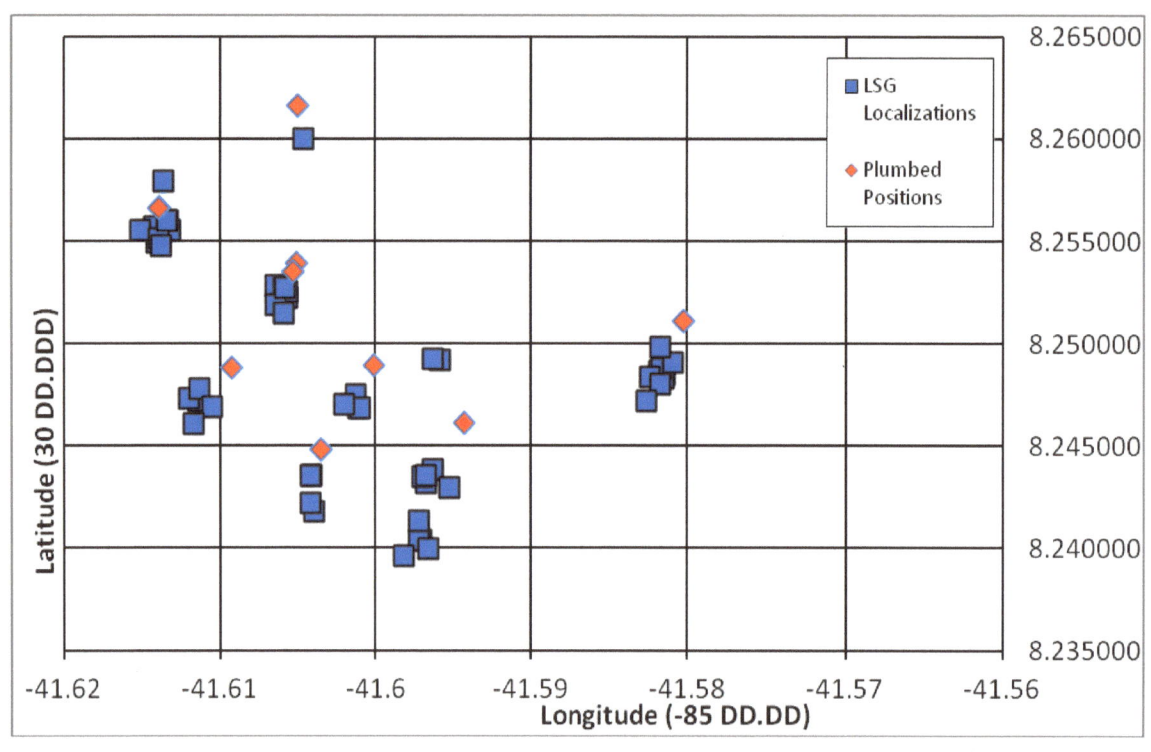

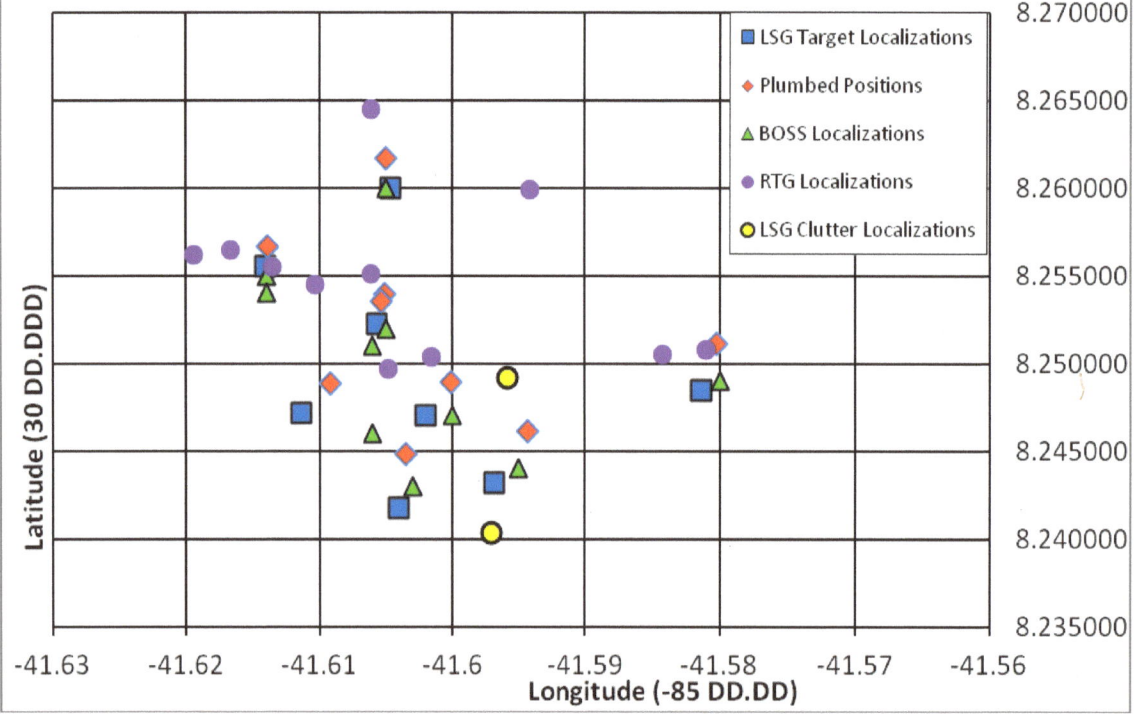

Figure 43. (*Top*) High confidence (CN>9) LSG Localizations. The furthest North target is only medium confidence (CN=8.02) due to the target being outside the survey.
(*Bottom*) Localization of RTG targets compared to LSG and BOSS localizations. The LSG targets shown represent the highest confidence from each cluster in the Top Figure.

7.7 Performance Objectives – Real-time Tracking Gradiometer (RTG)

The performance objectives for the RTG system are summarized in Table 10. The objectives include demonstration of the performance of the probability of detection, measurement of magnetic moment and target localization accuracy against proud and fully buried UXO targets.

Table 10. Performance Objectives for RTG Sensor

Performance Objective	Metric	Data Required	Success Criteria
Quantitative Performance Objectives			
Detection of all munitions of interest	Percent detected of seeded items	• Location of seeded Items	Pd=0.90
Measurement of ferrous target magnetic moment	For individual planted targets, the percent of targets whose magnetic moment is measured during a flyover to within 20% of that measured in the laboratory, for the particular altitude flown.	• Validation data for 15 selected targets	Demonstration of >90% of individual ferrous targets moments measured to within 20% of actual.
Location accuracy	Average localization error in meters from latitude and longitude ground truth localization for seed items	• Location of seed items surveyed to accuracy of 1m	ΔLocalization Error < 1m from ground truth
Qualitative Performance Objectives			
Ease of use		• Feedback from technician on usability of technology and time required	

7.7.1 Objective: Detection of All Munitions of Interest - Real-time Tracking Gradiometer (RTG)

The effectiveness of the RTG sensor in detecting proud to fully buried small, magnetic UXO targets and discriminating targets that are clustered together.

7.7.1.1 – Metric

A total of fourteen UXO targets were placed. These were comprised of ten class M objects and four class K objects. Five of these were individual targets (3 K-class and 2 M-class). Three were clusters, configured as 4-M targets, 2-M targets, and a combination of 2-M plus 1-K targets, respectively. Figure 12 illustrates the layout of the target field and Table 3 provides a listing of the targets' plumbed positions and descriptions.

7.7.1.2 – Data Requirements

The RTG data analysis demonstrated classification of 3 out of 5 individual targets and all 3 target clusters. Because of limitations in the magnetic signal processing, the RTG is not currently suitable for separating targets closely spaced in a cluster.

The LSG classified 8 groupings (5 individual targets and 3 cluster targets), or 100% of the groupings. Again, the LSG signal processing is not currently suitable for separating targets closely spaced in a cluster.

7.7.1.3 – Success Criteria

The RTG classified and localized 3 of 5 individual targets and all 3 target clusters other than the 50 caliber shells with low operator confidence resulting in a probability of classification of 0.66. The objective to classify the targets with a probability of 0.95 or above was not met. However it should be noted that the resolution of the RTG is such that it is, for the most part, unable to discriminate the closely packed targets within the clusters. If looking solely at cluster classification, which is almost as useful as individual target classification, the RTG was able to locate 3 of 3 clusters resulting in a probability of classification of 1.0

The LSG classified and localized all 5 isolated targets and all 3 target clusters with high operator confidence resulting in a probability of classification of 1.0. Of the 3 clusters of targets containing M and K class targets, the LSG was able to find all 3 clusters, as well as two ferrous clutter objects that were not detected by other sensors. If concerned with individual UXOs in a cluster, the objective was not strictly met. Again however, the LSG's resolution is such that it is almost impossible to discriminate the individual items within the clusters. It is also worth noting that the magnetic dipole strength of the clumps of 50 caliber shells is small enough as to be undetectable by either system, even if operating at a lower altitude over bottom.

7.7.2 Objective: Measurement of ferrous target magnetic moment - Real-time Tracking Gradiometer (RTG)

This objective focuses on determining the effectiveness of the RTG sensor in discriminating targets of interest and non-targets with high confidence. However, due to that all targets seeded in the field are of interest, what we present in this section is the capability of the RTG and LSG to discriminate between the UXO target types.

7.7.2.1 – Metric

The metric used to asses this objective is to identify the number of false targets eliminated at a specified confidence level. However, because the seeded field does not have targets of non-interest or planned clutter, the metric used is the number of targets correctly correlated with the target type seeded at the plumbed position, along with accurate estimation of the target dipole strength. As no measurements have been done to accurately determine the actual target dipole strength, this is estimation only.

7.7.2.2 – Data Requirements

Three types of cylindrical-shaped UXO targets were seeded: class M targets with typical length and width of 0.68 and 0.16 meters, respectively, class K targets with typical length and width of 1.64 and 0.27 meters, respectively, were used. Based on the size and composition of the various targets, we can estimate the target dipoles to be less than 50 A-m^2 (clusters will look larger).

7.7.2.3 – Success Criteria

While the target localization had high error, the dipoles calculated fall within the desired range, with the smallest detection at 13 A-m^2 and the largest at 80 A-m^2. However, due to the poor localization and the number of targets detected by the RTG, we cannot consider this objective successful.

However the LSG's results are consistent with the dipole strength estimations, allowing accurate discrimination. For the LSG the objective was met.

The magnitude of the dipole strength is useful in this context for several reasons. Accurate measurements can lead to the discrimination of different types of UXO from each other as well as from possible ferrous clutter. Dipole strength can also be indicative of a cluster of targets rather than an individual piece of ordinance.

7.7.3 Objective: Location Accuracy- Real-time Tracking Gradiometer (RTG)

This objective focuses on the effectiveness of the RTG sensor for localization of the UXO targets.

7.7.3.1 – Metric

The metric to assess this objective is to compare the RTG and LSG target localization with the diver provided plumbed positions.

7.7.3.2 – Data Requirements

RTG data was collected over the seeded field and analysis followed to detect and localize the targets of interest. The RTG localization for each target was cataloged and verified against the target listing and the plumbed positions provided by the divers.

7.7.3.3 – Success Criteria

Given the layout of RTG localizations as shown in Figure 43 it is clear this objective was not met. Of the targets seen, only a handful of the detections can with any confidence be directly related to a specific target, unfortunately making any average distance to plumbed location meaningless. Furthermore, refer to Section 7.4.3.3 for a discussion of sources of error that would lead to this objective not being met; even assuming the sensor was fully operational.

For the LSG however, the objective was successfully met here as well. A roughly 5m offset was apparent between the two vehicles' navigation, differences which are to be expected between separate runs by different vehicles. Correcting for this uniform 5m shift due to navigation error, the results shown in Table 9 and Figure 43 are separated from the plumbed positions by only 1-2m, which although not meeting the objective, are within the expected bounds as discussed before.

7.7.4 Objective: Ease of Use - Real-time Tracking Gradiometer (RTG)

This objective provides operator feedback on the ease of use of the RTG and its data PMA efforts.

7.7.4.1 – Data Requirements

The installation of the RTG primary components (sensor head and electronics bottle) onto the Bluefin12 UUV is straightforward and takes approximately 15 minutes to complete. Pre-mission testing usually takes a couple of hours which includes deploying the UUV in the water for ballasting and testing all sensors. Longer pre-mission setup can occur if ground faults are detected, requiring vehicle recovery back to the laboratory, opening the vehicle, and dealing with the supposed watertight cabling between the sensor and the electronics.

The UUV operator programs the state (power on and record, power on and standby, power off) of all sensors when planning survey missions while data quality can only be confirmed during PMA.

The RTG PMA effort, when dealing with the fully operational sensor, is a fairly straightforward two-step process: 1) upon recovery of the UUV after mission completion, all data is downloaded via Ethernet connection and 2) PMA follows once data has been downloaded. Normally, the data analyst will set up the RTG PMA algorithms to automatically sequence through the data files and perform the dipole search while cataloguing all of the results. With multiple channels malfunctioning, the PMA process was more complicated, involving altering the algorithms to search using only 4 gradients, properly devaluing the one new gradient that was used for some runs, and attempting to recalibrate the sensor data in order to reduce residuals.

The LSG PMA algorithms are similar to those of the RTG. In addition, an embedded version of these algorithms for the LSG has demonstrated results very consistent with the PMA version.

7.8 Summary – Real-time Tracking Gradiometer (RTG)

When in a fully operational state, the RTG is an invaluable tool in localizing ferrous targets of interest and when combined with the BOSS system, can aid discrimination between ferrous and non-ferrous targets of interest. However, due to unfortunate hardware issues, the RTG was not operating at normal efficiency. This led to reduced detections, poor localizations, and low confidence in the results. As with the BOSS system, improvements in vehicle navigation and ground truth accuracy would be beneficial in reducing errors; however a more pressing concern for the RTG is repair and refurbishment of the aging sensor system that doesn't have financial backing.

The LSG has the advantage of being an intrinsically more sensitive system than the RTG, making it more capable of detecting ferrous targets, measuring the dipole moment vector, and determining the location of ferrous targets of interest. Additionally, the LSG is less prone to failures than the RTG because it is still being supported by the developmental contractor, while the RTG is not. Although it is currently located on a separate vehicle (REMUS 600), requiring additional operators and introducing additional navigation errors, the LSG system makes a suitable replacement for the RTG in this context. In further consideration, the integration of the BOSS and the LSG sensors together onto a single UUV would be a valuable upgrade for enhanced capability for which there has not been sponsorship to date.

In addition to the LSG sensor, on board the REMUS 600 UUV, there was a Marine Sonics dual-frequency (900/1800 kHz) sidescan sonar and an EO sensor capable of providing target imagery. At the time of the LSG survey, the water clarity was favorable for target imagery to be collected. Figure 44 shows the results.

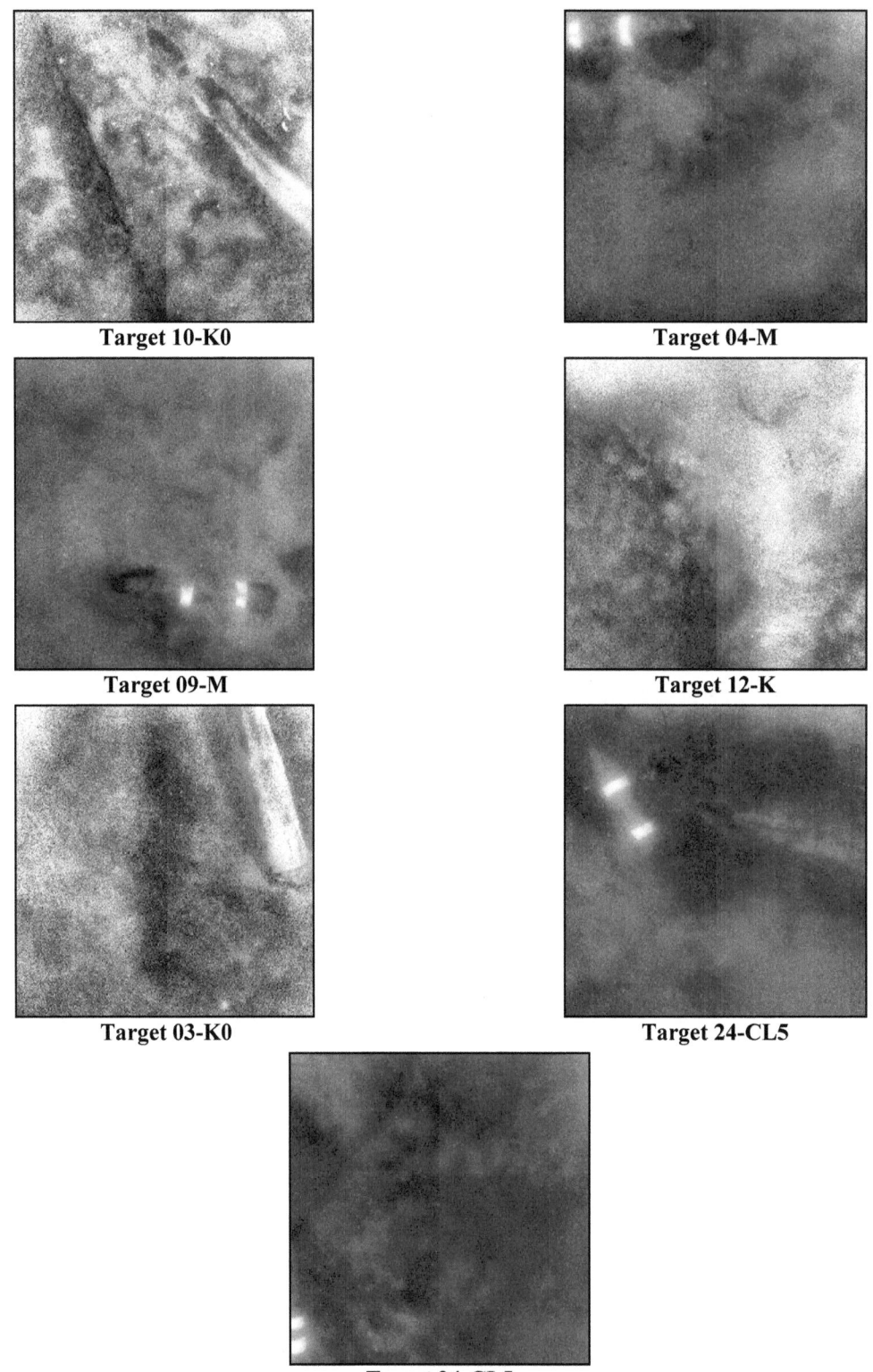

Target 10-K0

Target 04-M

Target 09-M

Target 12-K

Target 03-K0

Target 24-CL5

Target 24-CL5

Figure 44. Target imagery from the EO sensor on board the REMUS 600 UUV during the LSG survey

7.9 Value of Sensors

The fact that the BOSS sensor can image both proud and subsurface contacts suggests that the value of the REMUS 100 with the 900/1800 KHz sidescan sonar may have less value. However, since the REMUS 100 system is inexpensive to operate and provides a high area coverage rate, it provides an effective pre-survey of an area to assist in planning subsequent surveying of the area. The pre-survey can establish the provisional perimeter of the field, identify obstacles to UUV navigation at lower altitudes and identify features suitable for geo-referencing the survey. In addition, the REMUS 100's high-resolution imaging sonar (e.g., real-aperture sidescan, SAS or acoustic camera) offers high quality images of proud and partially buried targets, which is useful when optical visibility is low, and until such time that a "BOSS" upgrade becomes available with comparable imaging fidelity.

The ability of the BOSS system to image not only proud but fully buried targets ranging in size down to the M-targets and smaller provides a capability to assess the established field for fully buried targets. The LSG or RTG magnetic sensors allow for the discrimination of ferrous and non-ferrous targets and fusion of this data with sonar imagery significantly reduces false alarms. The electro-optic sensor provides images of proud contacts and can help in characterization of the sediment surrounding the contact.

The BOSS sonar should be considered the key, or primary sensor of the suite of sensors tested, because it can detect both proud and buried contacts and can provide the measurement of target dimensions and image features (a potential identification capability). In addition, the BOSS provides a measurement of burial depth. The magnetic gradiometer, in this case the LSG, is essential as a secondary sensor in that it provides confirmation that contacts are ferrous or non-ferrous. The electro-optic sensor provides images of proud contacts. Data products shown in figure 45 include a BOSS 3D image (left), Marine Sonics 1800-kHz Acoustic image (Middle) and Optical image (Right) collected with the REMUS 600 BMI System. The LSG contact positions from multiple tracks are displayed by white circles overlaid on the sonar images.

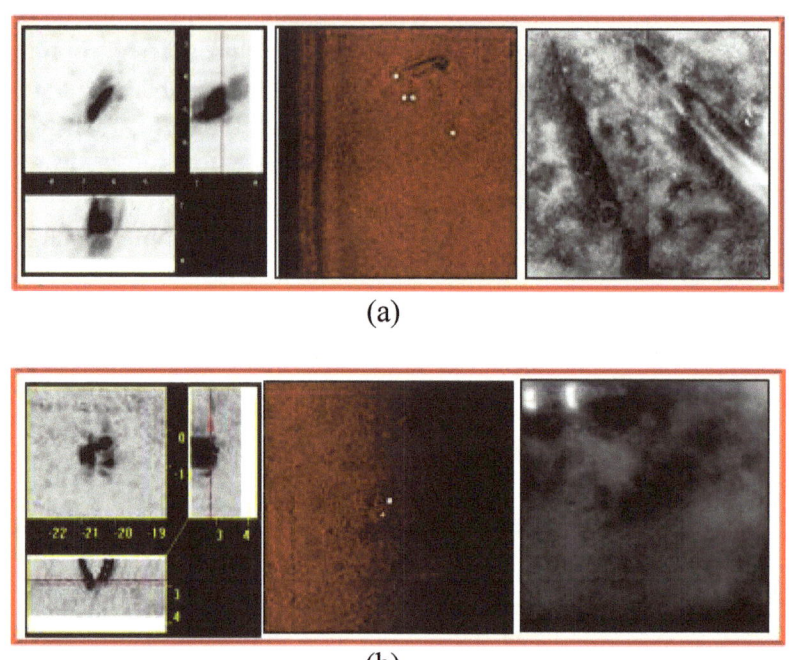

(a)

(b)

Figure 45. Product showing combined data from all sensors: (a) data from target 10-K0, (b) data from target 04-M. The LSG contact positions from multiple tracks are displayed by white circles overlaid on the sonar images. In the BOSS zoom images, the vertical axes are cross track, and the horizontal axes are along track, in meters.

8.0 COST ASSESSMENT

8.1 COST MODEL

Table 11 shows the cost model for the UUV-based survey system using the BOSS and RTG sensor system.

Table 11. Cost Model for UUV-based Survey System using BOSS and RTG

Cost Element	Data to be Tracked	Estimated Costs
Pre-test Preparations	• Assemble and check out system • System assembly • Checkouts of the vehicle and sensors • In water checks of the vehicle and sensors	$10-15K/day
Survey Planning	• Cost to review area maps, consider prevailing currents, consider launch facilities and interim transits, and plan and pre-program the UUV search tracks for the survey mission • System shipping costs • Installing Launch & Recovery Devise (LRD) on boat • Support Personnel	$1K/Day – Cost to Review $5-10K-Shipping $10-15K- Installing LRD $5-7K/Day – Support Personnel
Perform Survey	• Cost to conduct the test field survey • Boat support • Personnel support • Planned operational survey time • Allowance for weather and downtime	$20-25K/Day
Data Processing Costs	• Cost to conduct data analysis • Resources (Computer time) • Personnel hours	$5K/Day
Demobilization	• Cost for maintenance, breakdown, and packing • Personnel hours	$10-20K

A description of the cost elements in table 11 follows:

8.1.1 Pre-test preparations

Description: The BMI sensor suite and the Bluefin UUV are complex system and require careful assembly and checkout to assure correct operation in the field. The instruments are contained in separate sections that are individually packed for storage and must be assembled to the vehicle.

Supporting analysis / data: Actual schedule time and cost records from the Corporate (financial) Database at NSWC PCD, plus recall of schedule time from personnel actually working on the project during the experiment setup.

Interpretation of data: Test cost preparation time noted for the experimental effort would be approximately accurate in representing time required by a contract effort at a munitions site. Costs associated with labor would be expected to be drop over repeated missions, based on increased efficiency from experience gained over time and the potential for streamlining processes and equipment. Multiple systems would allow for exchange of parts to keep systems running in event of component failures.

8.1.2 Survey Planning

Description: Survey Planning includes initial study of the area of interest to determine the best way to perform the survey, considering the known environmental factors in the area. Cost of shipping equipment, both to and from the site, is considered here because it is specific to the geography of the survey area.

Supporting analysis / data: An estimate of labor required to research a site using available maps and NOAA data will be used. This includes researching available boat services, and using actual costs for rigging services to install the LRD onto boats.

Interpretation of data: Some variation in costs is expected depending on the area of interest and its remoteness. Areas with access to roads, docks, and boat/crane services will reduce the costs to plan, and later implement the survey.

8.1.3 Perform Survey

Description: Perform Survey estimates costs for boat services and personnel that support the actual survey, and also includes additional costs to account for possible weather and failure-related down time. During down time personnel must be kept available to make repairs or to restart operations when conditions allow.

Supporting analysis / data: A daily labor cost will be used based on the minimum required operating crew for the BMI sensor suite and UUV. Contract boat costs based on historical costs at Panama City will also be used.

Interpretation of data: This data should accurately reflect the costs of performing a survey in a munitions site.

8.1.4 Data Processing Costs

Description: Data Processing Costs are primarily labor to process and analyze data from the BOSS and RTG sensors, along with navigation and inertial data from the Bluefin 12 UUV to map and characterize detected munitions.

Supporting analysis / data: The hourly costs for NSWC PCD an engineer and the calculated cost for hours per square meter of surveyed field will be used. The processing time per area field surveyed will be developed based on actual time and area from the experimental effort.

Interpretation of data: This data should accurately reflect the costs of performing a survey in a munitions site.

8.1.5 Demobilization

Description: Demobilization costs are labor and material associated with breaking down, cleaning, and repacking the equipment for future use.

Supporting analysis / data: Labor time for demobilization from the experimental effort will be used based.

Interpretation of data: This supporting analysis data should accurately reflect the costs of demobilization in an actual survey.

8.2 Cost Drivers

Primary cost drivers for the system include maintenance of the systems , and the fact that the system as a whole is unique and still experimental would drive the cost to allowing turnkey analysis by users who are less experienced with the system.

8.3 Cost Benefits

A major cost benefit of the BMI sensor suite is that two sensing modalities are carried on one unmanned system to survey using two separate methods at one time. Acoustically reflective objects can be classified as ferrous or non-ferrous, and their size, aspect, and depth can be estimated in one pass. This reduces survey time and minimizes the problem of co-registering data from multiple sensors carried separately.

9.0 IMPLEMENTATION ISSUES

Regulatory issues are minimal, as the system does not release any materials or move any material in the environment. The acoustic radiation emitted by the system is not known to harm any marine organisms. Regulations regarding waterway traffic apply as with any craft, and local authorities will have to be engaged when releasing a UUV in public waters. Environment Mitigation / Protective Measures are covered in Appendix B.

The problems during the test with the RTG sensor are the major concern regarding using this system for munitions surveys. However, the failure of multiple channels in this sensor can be corrected. We recommend that repairs be made to the sensor using funds that were saved by leveraging test facilities made available during the 2011 ONR MCM S&T Demonstration.

The sensor suite used in the experiment is unique, research and development equipment developed by the Navy for finding buried mines. As such, it cannot be procured commercially. Additional equipment would have to be custom-built.

Improvements in magnetic sensors have been made since the RTG sensor was constructed, and alternate sensor designs might be considered, including the LSG sensor discussed earlier in this report. Collaboration between government and contractor subject matter experts would be required to develop a state of the art gradiometer suitable for integration with the Bluefin12 UUV and the BOSS sensor for regular use in munitions surveying. As a part of that effort, software for processing gradiometer and BOSS data would be streamlined to enable survey contractors to use the system with minimal additional training.

10.0 REFERENCES

**National Unmanned Systems Shared Resource Center (NUSSRC) Test Plan
18 March 2011 / Revision No: 1.1**

APPENDICES

Appendix A: Points of Contact

Team Members	Organization	Phone, Fax, E-mail	Function
Robert Leasko	NSWC PCD	Ph:(850) 235-5089 Fax:(850)234-4141 robert.leasko@navy.mil	Project Oversight
Charles Bernstein	NSWC PCD	Ph:(850) 234-4083 Fax:(850)234-4141 charles.bernstein@navy.mil	System Engineer
Amanda Mackintosh	NSWC PCD	Ph: (850) 230-7438 Amanda.mackintosh@navy.mil	Test Director
Ana Ziegler	NSWC PCD	Ph:(850) 636-6062 Fax:(850)234-4141 ana.ziegler@navy.mil	UUV Operator
Richard Holtzapple	NSWC PCD	Ph:(850) 234-4655 richard.holtzapple@navy.mil	BOSS Sensor Specialist
Jesse Angle, PhD	NSWC PCD	Ph:(850) 636-6458 jesse.angle@navy.mil	RTG Sensor Specialist

Appendix B: Environment Mitigation / Protective Measures

National Unmanned Systems Shared Resource Center Mitigation/Protective Measures

Mitigation/protective measures to be implemented during the NUSSRC test activities are described below. These measures are consistent with those identified in references (a) through (e) and standard U.S. Navy practice. These measures will minimize any potential effects to marine mammals or threatened and endangered species. Marine Species Observers will be required when operations exceed the parameters listed in the following table:

System Name	Operational Time Limit (Hours)
BOSS	< 6
LSG	< 15
SSAM	< 1.5
SSAM(2)	< 6
SAS12	< 6
RELIANT	< 15
BPAUV	< 15
REMUS	< 15
KLEIN	No Limit
SEAFOX	< 15
ACHERFISH	< 15
SEALION	< 15
RANGER	< 15
TRANSPHIBIAN	< 15
RANGER XN	< 15
IVERS	< 15
ACOUSTIC CAMERA	No Limit
SPIRAL WAVEFRONT TRACKING	< 15

Mitigation Measures -Sonar Operations
1. Personnel Training

a. All marine species observers onboard platforms involved in Naval Surface Warfare Center Panama City Division Research Development Test and Evaluation (RDT&E) activities shall complete Marine Species Awareness Training (MSAT).

b. Marine species observers shall be trained in the most effective means to ensure quick and effective communication within the command structure in order to facilitate implementation of mitigation measures, if marine species are spotted.

2. Marine Species Observer Responsibilities

a. On the bridge of surface vessels, there shall always be at least one to three marine species awareness trained observer(s) on watch, whose duties include observing the water surface around the vessel.

(1) For vessels with length less than 65 feet (20 meters), there shall always be at least one marine species observer on watch.

(2) For vessels with length between 65-200 feet (20-61 meters), there shall always be at least two marine species observers on watch.

(3) For vessels with length above 200 feet (61 meters), there shall always be at least three marine species observers on watch.

b. Each marine species observer shall have at their disposal at least one set of binoculars available to aid in the detection or marine mammals.

c. On surface vessels equipped with AN/SQQ-53C/56, pedestal mounted "Big Eyes" (20x110) binoculars shall be present and in good working order to assist in the detection of marine mammals in the vicinity of the vessel.

d. Marine species observers shall employ visual search procedures employing a scanning methodology in accordance with the Lookout Training Handbook (NAVEDTRA 12968-D).

e. Marine species observers shall scan the water from the vessel to the horizon and be responsible for ensuring that all contacts in their sector follow the below protocols:

(1) In searching the assigned sector, the marine species observer shall always start at the forward part of the sector and search aft (toward the back).

(2) To search and scan, the marine species observer shall hold the binoculars steady so the horizon is in the top third of the field of vision and direct the eyes just below the horizon.

(3) The marine species observer shall scan for approximately five seconds in as many small steps as possible across the field seen through the binoculars.

(4) The marine species observer shall search the entire sector in approximately five-degree steps, pausing between steps for approximately five seconds to scan the field of view.

(5) At the end of the sector search, the glasses will be lowered to allow the eyes to rest for a few seconds, and then the marine observer shall search back across the sector with the naked eye.

f. After sunset and prior to sunrise, marine species observers shall employ Night Lookout Techniques in accordance with the Lookout Training Handbook.

g. At night, marine species observers shall scan the horizon in a series of movements that will allow their eyes to come to periodic rests as they scan the sectors. When visually searching at night, marine species observers shall look a little to one side, and out of the corners of their eyes paying attention to the things on the outer edges of their field of vision.

h. Marine species observers shall be responsible for reporting all objects or anomalies sighted in the water (regardless of the distance from the vessel) to the Test Director or the Test Director's designee.

3. Operating Procedures

a. The Test Director or the Test Director's designee shall maintain the logs and records documenting RDT&E activities should they be required for event reconstruction purposes. Logs and records will be kept for a period of 30 days following completion of a RDT&E mission activity.

b. A Record of Environmental Consideration shall be included in the test plan prior to the test event to further disseminate the personnel testing requirement and general marine mammal mitigation measures.

c. Test Directors shall make use of marine species detection cures and information to limit interaction with marine species to the maximum extent possible consistent with safety of the vessel.

d. All personnel engaged in passive acoustic sonar operations (including aircraft or surface vessels) shall monitor for marine mammal vocalizations and report the detection of any marine mammal to the Test Director or the Test Director's designee for dissemination and appropriate action.

e. During High Frequency Active Sonar/Mid Frequency Active Sonar mission activities, personnel shall utilize all available sensor and optical systems (such as night vision goggles) to aid in the detection of marine mammals.

f. Naval aircraft participating in RDT&E activities at sea shall conduct and maintain surveillance for marine species of concern, as long as it does not violate safety constraints or interfere with the accomplishment of primary operational duties.

g. Marine mammal detection shall be immediately reported to the Test Director or the Test Director's designee for further dissemination to vessels in the vicinity of the marine species as appropriate where it is reasonable to conclude that the course of the vessel will likely result in a closing of the distance to the detected marine mammal.

h. Safety Zones for tests that require the use of safety zones, when marine mammals are detected by any means (aircraft, shipboard lookout, or acoustically) within 914 meters (1,000 yards) of the sonar system, the platform will limit active transmission levels to at least six decibels below normal operating levels for those systems that this capability is available.

(1) Vessels will continue to limit maximum transmission levels by this six decibels factor for those systems that this capability is available until the animal has been seen to leave the area, has not been detected for 30 minutes, or the vessel has transited more than 1,829 meters (2,000 yards) beyond the location of the last detection.

(2) Should a marine mammal be detected within or closing to inside 457 meters (500 yards) of the sonar system, active sonar transmissions will be limited to at least 10 decibels below the equipment's normal operating level for those systems that this capability available. Platforms will continue to limit maximum ping levels by this 10 decibels factor until the animal has been seen to leave the area, has not been detected for 30 minutes, or the vessel has transited more than 1,829 meters (2,000 yards) beyond the location of the last detection.

(3) Should the marine mammal be detected within or closing to inside 183 meters (200 yards) of the sonar system, active sonar transmissions will cease. Sonar will not resume until the animal has been seen to leave the area, has not been detected for 30 minutes, or the vessel has transited more than 1,829 meters (2,000 yards) beyond the location of the last detection.

(4) If the need for power-down should arise, as detailed in paragraph 3h above, U.S. Navy staff will follow the requirements as though they were operating the normal

operating level of 235 decibels (i.e., the first power-down will be to 229 dB, regardless of the level above 235 dB the sonar was being operated) .

(5) Special Conditions Applicable for Bow-Riding Dolphins. If, after conducting an initial maneuver to avoid close quarters with dolphins, the ship concludes that dolphins are deliberately closing in on the ship to ride the vessel's bow wave, no further mitigation actions will be necessary because dolphins are out of the main transmission axis of the active sonar while in the shallow-wave area of the vessel bow.

i. Prior to start up or restart of active sonar, operators will check that the safety zone radius around the sound system is clear of marine mammals.

j. Sonar levels (generally). U.S. Navy will operate sonar at the lowest practicable level, not to exceed 235 decibels, except as required to meet RDT&E objectives.

k. These procedures will also apply as much as possible during Autonomous Underwater Vehicle/Unmanned Underwater Vehicle (AUV/UUV) operations. An observer will be located on the support vessel or platform to observe the area, depending on the test event. When an AUV/UUV is operating and shows potential to expose, it is impossible to follow and observe it during the entire path but they will visualize the general area or modify the plan specific to the nature of the system. If the system is undergoing a small track close to the support platform, then observers will be used.

m. Safe standoff distances for swimmers and divers are detailed in the U.S. Navy Dive Manual. These distances will be used as the standard sonar safety buffer for operations occurring within the NSWC PCD Study Area.

Mitigation Measures -Surface Operations

1. While underway, vessels shall have at least one to three Marine Species Awareness Trained (MSAT) observers (based on vessel length) with binoculars. As part of their regular duties, marine observers shall watch for and report to the Test Director or Test Director's designee the presence of marine mammals.

2. Marine observers shall employ visual search procedures employing a scanning method in accordance with the Lookout Training Handbook (NAVEDTRA 1296B-D).

3. While in transit, naval vessels shall be alert at all times, use extreme caution, and proceed at a "safe speed" (the minimum speed at which missions goals or safety will not be compromised), so that the vessel can take proper and effective action to avoid collision with any marine animal and can be stopped within a distance appropriate to the prevailing circumstances and conditions.

4. When marine mammals have been sighted in the area, naval vessels shall increase vigilance and shall implement measures to avoid collisions with marine mammals and avoid activities that might result in close interaction of naval assets and marine mammals. Actions shall include changing speed and/or direction and are dictated by environmental and other conditions (e.g., safety, weather).

5. Naval vessels shall maneuver to keep at least 460 meters (500 yards) away from any observed whale and avoid approaching whales head-on. This requirement does not apply if a vessel's safety is threatened, such as when change of course will create an imminent and serious threat to a person, vessel, or aircraft; and to the extent vessels are restricted in their ability to maneuver. Vessels shall take reasonable steps to alert other U.S. Navy vessels in the vicinity of the whale.

6. Where operationally feasible and safe, vessels shall avoid closing within 183 meters (200 yards) of marine mammals other than whales.

Protective Measures -Surface Operations

1. Surface vessels shall not operate over areas of sea grass. Marine vehicle operators shall observe idle speed limits, channel markers, and other aids to navigation to avoid any effects to nearby sea grass.

Protective Measures -Subsurface Operations

1. No mine-like object (MLO) and versatile exercise mine (VEM) placement or crawler operations will occur within areas of sea grass.

2. No MLO and VEM placement or crawler operations will occur within areas of known hard bottom.

3. Activities such as mine placement and crawler operations that cause bottom disturbance will not be conducted in Marine Managed Areas.

4. Mine placement and anchoring will not be conducted in areas that could damage hard bottom or sea grass habitats.

5. Activities, such as mine placement and crawler operations that cause bottom disturbance, will not occur over artificial reefs or known shipwrecks. If an unknown shipwreck is uncovered, the State Historic Preservation Officer will be notified and all activities will cease.

References:

a. NSWC PCD document: Mission Activities Environmental Impact Statement/Overseas Environmental Impact Statement
b. Assistant Secretary of the Navy (Installations and Environment) Record of Decision for NSWC PCD Mission Activities of 15 Jan 10
c. National Marine Fisheries Service Biological Opinion on 2010 Letter of Authorization for U.S. Navy Training in NSWC Panama City of 14 Jan 10
d. National Marine Fisheries Service Final Rule Taking and Importing Marine Mammals NSWC PCD Mission Activities of 21 Jan 10
e. NSWC PCD Marine Mammal Protection Act Letter of Authorization of 21 Jan 10

www.ingramcontent.com/pod-product-compliance
Lightning Source LLC
Chambersburg PA
CBHW050743180526

45159CB00003B/1334

9781500254827